W9-CNB-809

# ROUTE 128

# ROUTE 128

## Lessons from Boston's High-Tech Community

SUSAN ROSEGRANT

AND

DAVID R. LAMPE

*A Division of* HarperCollins*Publishers*

Library of Congress Cataloging-in-Publication Data
Rosegrant, Susan, 1954–
Route 128: lessons from Boston's high-tech community/
Susan Rosegrant and David R. Lampe.
p. cm.
Includes bibliographical references and index.
ISBN 0–465–04639–8
1. High technology industries—Massachusetts—Boston Metropolitan Area. 2. Computer industry—Massachusetts—Boston Metropolitan Area. I. Lampe, David (David Ralph) II. Title.
HC108.B65R67 1992
338.4′762′000974461—dc20 91–58600
CIP


A Division of *HarperCollins*Publishers
PRINTED IN THE UNITED STATES OF AMERICA
*Designed by Ellen S. Levine*
92 93 94 95 CC/HC 9 8 7 6 5 4 3 2 1

*To Joanna and Luisa*

*Miracles are propitious accidents,*
*the natural causes of which are too complicated*
*to be readily understood.*

—George Santayana

# Contents

# Preface

In 1985, we set out to write a book about the "Massachusetts Miracle"—the economic boom of the 1970s and 1980s fueled by high-tech industry. At the time, the Massachusetts economy was near the peak of its performance. Computer and software companies were flying high. New high-tech firms were spreading throughout the region: along the famous Route 128 corridor around Boston; all the way to Worcester in central Massachusetts; and even into southern New Hampshire. Along with California's Silicon Valley, the "Route 128 phenomenon" was the quintessential high-tech success story.

Our goal was to shed light on the underpinnings of this new brand of high-technology–based regional economy that seemed to hold such bright promise for Massachusetts, other areas of the nation, and the world. What were the real forces that made it happen? Certainly, it was as much the product of culture, history, and forces in the marketplace as of politics and planning. Was there anything special about Massachusetts or its history? Were there lessons for other regions that were inspired by the "Miracle" and hoped to replicate it?

Talking to entrepreneurs, academics, and policymakers, digging into the past, and sifting through the facts all took time. And during that time, the "Massachusetts Miracle" came crashing to earth. Under the spotlight of Governor Michael Dukakis's failed presidential campaign, the "Massachusetts Miracle" became a bad joke to many, and—to some—a symbol of false hope and betrayal.

But despite the downturn, we were convinced that something special *was* going on in the state that, if not miraculous, was certainly remarkable. Indeed, Massachusetts has been a fountainhead of high-tech innovation for more than a century, well before the existence of computer companies or even Route 128. And this underlying process of creating new ideas and encouraging their flow to the marketplace is still going on.

Our purpose became to put the so-called "Massachusetts Miracle" into a much broader context, to look beyond the state of the handful of computer companies by which the region is so often judged. We have focused instead on the long evolutionary process that laid the foundations for the entrepreneurial success stories—and failures—of Route 128.

One of the key lessons to be drawn from the region is to expect no miracles. Although new technologies may be at the root of economic success, "high tech" is not a magic wand that can conjure up something from nothing. Major breakthroughs and blockbuster companies are the exception, not the rule. The minicomputer industry notwithstanding, the vast majority of technological advances are incremental, and the vast majority of high-tech start-ups address relatively small markets.

Perhaps the most important lesson that emerges, however, is that the extraordinary concentration of high-tech innovation that has flourished in places like Massachusetts and California springs from deep within our American culture. It is a product of our unique systems of government, business, and higher education. More than any particular individual or organization, these sectors have provided the fertile soil for America's extraordinary record of technological and economic success.

Our nation has been lucky. Our universities, our style of government, and our entrepreneurial spirit are all national resources that have served us well by nurturing a highly flexible and creative environment for science and business. Unfortunately, we have taken these resources for granted, reaping the rewards without fully understanding how they have worked for us.

To sustain us through the economic challenges we face today in a more competitive world and to maintain our leadership in high technology, we need to nurture these uniquely American resources and to adapt them to different and continually changing conditions. Inevitably, this will involve learning from other nations and incorporating the best of what we see. But if we are to prevail, we must begin at home. We can succeed only if we understand ourselves.

# Acknowledgments

A WIDE RANGE OF PEOPLE contributed to the completion of this book. We are deeply indebted to the entrepreneurs, venture capitalists, economists, academics, government officials, and others who took the time to talk to us about their experiences in the Route 128 region, and to share their views on what it all means.

A significant portion of the work of writing this volume was done under the auspices of the Science, Technology, and Public Policy Program at the John F. Kennedy School of Government at Harvard University. We would like to thank Professor Lewis M. Branscomb, Director of the Program, and Professor Harvey Brooks, who reviewed and commented on early drafts.

Special thanks are due to Arthur L. Singer of the Alfred P. Sloan Foundation who saw the need for a broader view of the significance of the Route 128 region. Without the grant arranged through his efforts, this book would not have been possible. We would also like to thank Eric C. Johnson of MIT, who first suggested that we seek such support, and who continued to urge us on in his own particular manner. The interest and encouragement of Professor James M. Utterback and Thomas R. Moebus of MIT were also greatly appreciated.

We are profoundly grateful to Barbara Feinberg. Her cheerful encouragement, sage advice, valuable feedback, and top-notch editorial assistance were crucial to our ability to complete this project.

Our families have been extraordinarily understanding throughout the entire process. Their patience and unbending support gave us the strength to carry on through all of the difficulties that arose along the way. We also thank the teachers of the Concord Children's Center and Mary Lee Worthy, who, by providing the very best care for our children, allowed us to devote our energies to the task at hand.

Finally, special thanks go to Jane Rosegrant, Andrew Tweedy, Maria Pitaro, and Steve Condiotti for helping us to stay in touch with reality.

The interviews conducted for *Route 128* were done over a period from 1985 through 1991. Longer interviews with entrepreneurs were conducted in person whenever possible, and taped. A number were conducted over the telephone and taped, with the full knowledge of the person being interviewed. A large number of shorter interviews, conducted either in person or on the telephone, were recorded in note form. In all cases, the persons interviewed were informed in advance that the statements they made, and the facts they provided, would be published.

# 1

# In Search of the "Miracle"

THE DECADE ENDING IN 1988 was a dizzying time in Massachusetts. Beginning in the late 1970s, a dramatic boom in the state's economy propelled the Bay State into national and international prominence. Behind the boom lay the products and the promise of high technology: Massachusetts was touted as a shining example of how an aging industrial economy with declining industries and rising unemployment could pull itself up by its bootstraps, relying on new technology-based companies to chisel out niches in the increasingly competitive world marketplace. Many in the state were gripped by a sense of unlimited possibilities and seemingly endless growth.

To be sure, Massachusetts was not alone; other states—particularly in the Northeast, where much of the period's financial wheelings and dealings were concentrated—were experiencing a period of giddy growth. But in contrast to the paper profits rolling in elsewhere, Massachusetts had more tangible signs of prosperity—new products, new companies, and new jobs.

By the late 1980s, the state's nearly 3,000 high-tech companies had dramatically altered the look, the feel, the spirit, and the prospects of Massachusetts, particularly in the Greater Boston area. The impact of these companies could be felt everywhere. With the exception of California's Silicon Valley, no other region of similar size in the world had produced so many new high-tech companies, and in terms of diversity, Massachusetts was unsurpassed. From the perspective of the rest of the country, in some regions struggling to revitalize moribund, Rust Belt industries, the development of this mecca of future-oriented industry appeared miraculous. Indeed, a "Miracle" is what it was called.

The region's high visibility increased even more in 1988 when three-term Governor Michael Dukakis leapt into the national spotlight as the Democratic contender for the presidency, hoping to parlay the magic of Massachusetts into a place in the White House. In the excitement generated by the state's economic transformation, the presidential election, and the flood of media attention, a sort of mythology developed about the region. Even some normally reserved New Englanders began to believe in the "Massachusetts Miracle."

Things seemed to happen at a breathless pace. Only a couple of years separated many new technology companies from their humble basement beginnings to their first public stock offerings, turning their founders into instant millionaires. Moreover, new jobs were being created in a broad array of businesses as established companies expanded, new companies sprang up, and service and supply firms blossomed to meet their needs. Computer and software companies in the state were expanding so fast that the high-tech network pressured state schools to graduate more engineers. And overall unemployment had dropped so low that even the hard-core unemployed, including inner-city youth and recent immigrants, were reaping the benefits. Around the world, Route

128—a sixty-five-mile highway arcing around the cities of Boston and Cambridge, where many high-tech companies got their start—had become a symbol of what high tech could achieve.

Encircled by and radiating out from Route 128 were thousands of companies with names like Thinking Machines, Candela Laser, Stratus Computer, Genetics Institute, and Parametric Technology. The sophisticated technologies at the heart of these firms read like a dictionary of high-tech buzzwords: artificial intelligence; medical lasers; fault-tolerant computers; biotechnology; and software for computer-aided design.

Although the region did not bloom in the national consciousness until the 1980s, when politicians, economists, and journalists hailed these entrepreneurial firms as the force behind the Massachusetts Miracle, the proliferation of new technology-oriented firms in the area was winning modest attention as "the Route 128 phenomenon" as early as the 1950s. It was in the 1950s, after all, that Ken Olsen founded Digital Equipment Corporation—a company that perhaps more than any other came to represent the era. The story of how Olsen built DEC up from its humble beginnings has become part of the local folklore.

After completing his Ph.D. at the Massachusetts Institute of Technology (MIT) in 1950, Olsen was asked to join Project Whirlwind, a crack team of students and researchers that in 1953 built the world's first reliable digital electronic computer. With the Cold War in full swing, MIT was awash in federal money supporting a wide range of basic research that could help the United States keep the technological edge that had proved so crucial during World War II.

An intense, no-nonsense man with a strong practical bent, Olsen already had two patents on computer technology to his credit. Now, at thirty-one, he was the leader of a research group designing and building a new generation of computers that would replace with tiny transistors the bulky vacuum tubes then common in electronic equipment. The faster and more powerful new machines would be the cornerstone of the nation's sophisticated new air defense system.

But something was missing. "Nobody out there cared," he recalls. "In this case, we made transistorized computers, and the business world thought we were just academics. We published everything we knew, and it just had no impact."[1] Olsen confided his frustration to Harlan Anderson, one of his engineering colleagues at MIT. Olsen argued that the smaller computers they were working on had real potential for application in industry, and that the concepts they were developing were just too important to lie idle in the halls of academia. It was obvious, he insisted, that if the stodgy business world was ever going to sit up and take notice, they had to make it happen themselves. Anderson agreed and the two decided to launch a company.

Despite the booming postwar economy of the 1950s, and the optimism of the entrepreneurs, it was a risky prospect. At the time, commercial computers were million-dollar room-sized machines that required an army of specialists to operate and maintain. The only customers for the complex devices were huge government agencies, a few research universities, and a handful of large corporations. Most pundits predicted that the business community would never have much use for expensive machines that could grind through lots of calculations quickly, and *Fortune* magazine had recently claimed that no one was making any money in computers.

Undeterred, Olsen and Anderson checked out a pile of books on business plans from the Lexington town library, near where they lived. Together, they cooked up a proposal and took it to American Research and Development Corporation (AR&D), the pioneering local venture capital firm that had attracted widespread attention—and skepticism—for its novel practice of funding high-tech start-ups. Although AR&D's board had misgivings about the engineers' lack of business experience, it was intrigued with the new computer concept, which it had already caught wind of through its network of contacts. Insisting that Olsen and Anderson avoid using the word "computer" in the name of the proposed company to avoid scaring off potential investors, AR&D gave the entrepreneurs the go-ahead, backing the new Digital Equipment Corporation to the tune of $70,000.

Even in those days that wasn't a fortune, but it was enough. Olsen got his brother Stan to join them, and the three men set up shop above a furniture store on the second floor of a huge old textile factory in the faded mill town of Maynard, just 20 miles west of Boston. They rented 8,600 square feet of space for twenty-five cents a foot, including heat and the services of a night watchman. With Anderson's lawn chairs and an old rolltop desk for office furniture, the entrepreneurs rolled up their sleeves and went to work, doing everything from design to production themselves.

Three years after its launch, the fledgling firm produced a revolutionary product its founders called a "minicomputer." The machine's list price was a mere $120,000, and it took up no more room than a couple of big desks. Of particular appeal was the addition of a keyboard and video screen that allowed users to interact directly with the machine, correcting mistakes, resolving programming problems, and trying new approaches immediately. The friendly little computers were a hit. Over the next three decades, Olsen relied on the minicomputer concept to build Digital Equipment into the third largest computer company in the world.

By the 1980s, Olsen's Digital Equipment, by most measures the area's most successful post–World War II high-tech start-up, had become the symbol of the "new" class of company that could spring from basic research to breathe life into an ailing economy. What happened in Massachusetts captured the world's attention, but what seemed a miracle for some eventually proved a disaster for many. That the state experienced a sudden demand for its products—in particular minicomputers, software, weapons, and financial services—is unquestionable. The surge of growth that resulted from this demand was particularly visible, coming as it did after the difficult recession of the early 1970s. Massachusetts was anxious to embrace good times, and its enthusiasm knew no bounds.

But the very headiness of those days led to an era of speculation that contributed to the state's subsequent fall. Entrepreneurs dreamed up companies at a record rate, and easily found the money to start them; real estate developers offered new condominium and office complexes at record prices, and easily found people and businesses to fill them; even homeowners got into the act, buying and selling houses in a frenzied bid to move up and get ahead. Although Massachusetts residents were not alone in embracing the widespread 1980s values of consumerism and upward mobility, many faced real opportunities for sudden gain and acted on them.

It is significant that companies launched in the 1980s often pursued very different paths in their quest for success than the technology start-ups of a few decades earlier. The 1983 founding of Encore Computer, one of the East Coast's uncharacteristically flamboyant deals, is a particularly apt example of the gambling mentality of this period. In sharp contrast to DEC's modest beginnings, Encore's superstar founding team—consisting of former Prime Computer president Ken Fisher, former DEC technical guru Gordon Bell, and Data General cofounder Henry Burkhardt—was able to generate publicity, enthusiasm, and money solely on the basis of its reputation.

Encore made its debut, fittingly enough, at the posh Helmsley Palace Hotel in New York City. Although details on the precise nature of the company's proposed business were sketchy, journalists and investors swarmed around the startup. Before ever shipping a product, Encore had been featured in newspapers and magazines across the country and had raised $47 million in venture capital and an initial public offering.

In this intoxicating environment of "easy" wealth and expanding marketplaces, few people—whether bankers, executives, or developers—took the time to question strategies or challenge goals or assumptions. Again, Encore is a case in point. Despite—or even, perversely, because of—its generous influx of capital, Encore had a rocky start. Struggling to find an appropriate niche, the company revamped its original strategy—to serve as a kind of umbrella for a number of technology companies and products—to focus on manufacturing its own high-performance computers. Encore was continuously plagued by product delays and management turmoil, including the defections of Burkhardt and Bell. Eventually, the company merged with the much larger computer operations of Gould Incorporated and relocated to Fort Lauderdale, Florida. Gould, owned by Japanese Nippon Mining Company, got a controlling stake in Encore in the deal.

Although Encore's story may be extreme, it is just one example of the mentality of this period. Increasingly, the so-called "Miracle" was built on the back of unchecked expansion, false expectations, and unwarranted speculation. This shaky foundation may have held up during the go-go years of the early 1980s, but it could not cope with the overcrowded markets, increased global competition, shifts in technology, and the national economic downturn later in the decade.

The unraveling of the "Miracle" began as early as 1984 in some product areas, as venture capitalists and entrepreneurs discovered that their shared enthusiasm for new technology start-ups had created too many companies chasing the same markets. On top of that, the Route 128 region's mainstay minicomputer makers began to feel the effects of a wholesale switch in user demand from their midsized machines to smaller and less expensive personal computers and interconnected workstations.

Events from outside the region began to compound these earlier problems. The 1987 stock market crash and the ensuing national recession hit the entire Northeast hard, but Massachusetts seemed particularly vulnerable. In 1988, as the state's high-tech, financial services, and defense markets all weakened, Governor Dukakis lost his bid for the presidency, and the state logged its first budget deficit in thirteen years.

By 1989, the "Miracle" was miraculous no more. The formerly booming real estate market in Massachusetts crashed; unemployment began to rise; and layoffs, restructurings, acquisitions, and bankruptcies substantially altered the appearance and outlook of the high-tech community. A downward spiral had begun, and in 1992 the state was still struggling to fight its way back onto its feet.

## THE PROMISE OF HIGH TECHNOLOGY

With the benefit of hindsight, it is easy to see that the last decade in Massachusetts had a lot more to do with money—either the profusion or the lack of it—than with miracles. It can be harder to see beyond this to the *value* of what happened here. Both the hype of the boom years and the hysteria that accompanied the state's subsequent economic decline have led to some serious misconceptions about Massachusetts's strengths and value to the nation as a whole. The high-technology community, in particular, has disappointed many observers by not proving invulnerable to economic stagnation and by failing to deliver on its apparent early promise of consistent explosive growth and high returns on investment.

Many observers have cited the formation of new high-technology businesses as the essence of the Route 128 phenomenon, but the firms are just one outward manifestation of a more complex process that produces breakthroughs in the world of fundamental science and engineering and transfers them to the world of business.

During the 1980s, as articles in the national and international press extolled the virtues of high technology in stimulating economic development, visitors from all over the world seeking to bolster their own economies flocked to Massachusetts, looking for answers to their particular problems. During this period many observers began to equate the performance of the high-tech community with the state's overall economic and financial health.

This is a dangerous assumption. Economies, industries, and technologies all go in cycles. A strong high-tech community such as that in the Boston area does not guarantee a booming regional economy. It has, however, increased the likelihood of discovering, developing, and cashing in on emerging technologies in areas such as radio tubes, computers, software, or biotechnology. Historically, it has proven devilishly difficult to predict the real market potential of a new technology, or even to guess how long it will take before it reaches the marketplace. The rules of the innovation game appear to change with each new technology, and even with the passage of time.

Instead of focusing on short-term issues—whether a state has a deficit one year or not, or whether the computer industry happens to be laying off workers during a particular year—it is more important to look at the underlying process that allows the conception and application of a broad range of new ideas. In this complex and inherently long-term process the Boston area has excelled. And it is a phenomenon that has been a part of the region's character since long before the Massachusetts Miracle, even before Route 128, and before World War II.

The state's high-tech network has received more than its share of credit and blame for the shifting fortunes of the state's economy. Here are some of the things that Massachusetts's high-tech community is *not:*

- It is not all computer companies. The size and high visibility of such computer manufacturers as DEC, Data General, Prime, Wang, Apollo, Stratus, and others have led many observers to conclude that the computer business is the soul of the Route 128 region. Yet one of the key strengths of the Boston area—and one that sets it apart from other regions of the country—is the broad diversity of high-tech companies that call the area home.
- It is not the sole determinant of the region's economic health. Despite the intense scrutiny that has been turned on

Route 128, high-tech businesses actually are responsible for only about 10 percent of total employment in the state. Although this is still about three times the national average, it leaves 90 percent of the pie for other kinds of businesses.

- It is not typified by big companies. Industry giants like Raytheon, Lotus, and DEC, and smaller, highly visible companies like Analog Devices and Biogen tend to capture the headlines, but most high-tech companies are small firms with modest returns. Industry estimates hold that more than 60 percent of all high-tech manufacturing concerns in the state had annual sales of $5 million or less, and only about 20 percent of software firms topped $5 million.
- It is not replacing the manufacturing base. Manufacturing in the Northeast has declined slowly since the 1920s. The boom in computers in the 1970s temporarily slowed the loss of manufacturing jobs, but the erosion has since continued. The high-tech sector has contributed primarily to the creation of new service jobs. Major manufacturers in the region continue to start up new operations in more favorable locations in the United States and other countries.
- Its riches have not spread equitably throughout Massachusetts. Although the state government during the boom attempted with some success to use high tech to prop up ailing regional economies distant from Boston, more than half of the state's business activity takes place within twenty-five miles of the Hub—precisely the Route 128 region.
- It is not invulnerable. All sectors of the economy must weather cyclical downturns along with the good times. Massachusetts's high-tech community is no exception. In the early 1970s, for example, when a nationwide recession sent the state's economy reeling, there were widespread reports of engineers and scientists with Ph.D.s driving taxicabs in Boston.

Ten years later, companies were bemoaning the lack of qualified scientific and technical talent. At the beginning of the 1990s, a downward cycle was well established again.

• It is not dying. In 1991, Massachusetts businesses and residents had plenty to complain about, from a falling demand for key technological products, leading to company failures and massive layoffs, to the collapse of the real estate market. Despite these hardships, new high-tech businesses started up that year in the state. Innovation has continued as it has for decade upon decade in the region.

• It was not planned. Politicians, business leaders, university administrators, and even real estate developers have sought to take credit for various aspects of the Route 128 phenomenon at one time or another. But no individual, organization, or institution can lay claim to having the foresight to create this high-tech enclave. Few observers on the local scene had an inkling of the potential for this new business economy until it was already in full swing in the early 1980s.

## THE REAL MIRACLE

The economic surge that accompanied the phenomenal growth of the high-tech community in Massachusetts during the 1980s was not a miracle. Considering the state's difficulties in picking up the pieces after this period of inflated expectations, the so-called miracle years look more like a setup for a fall.

But while there may have been no miracle decade, the region's windfall of prosperity was an outward manifestation of a process that truly can be called miraculous—a process with a venerable history that has continued unabated during the current recession. Simply put, the real miracle of Massachusetts has been its exceptional ability—throughout time—to grow ideas.

The area that has come to be symbolized by Route 128 is one of the world's leading centers of innovation. It is a region where, on a grand scale, a broad array of emerging technologies has been developed and gotten its first taste of the marketplace. This is an unpredictable process that is not always dramatic, not always financially rewarding, and not always successful, but it is crucial to the productivity and competitiveness of the nation. A special blend of attitudes, traditions, and institutions in eastern Massachusetts has fostered an environment that not only stimulates the conception of novel technologies, but also nurtures their development and encourages their application, easing their way into the marketplace as new products, new processes, and even entirely new industries.

Innovation is not confined to any particular geographic area of the country, nor to any particular class of company, but the Boston area has been at the leading edge of developments in a broad range of industrially important new technologies for nearly two centuries. No other region can match this record of continued productivity.

What is it then that makes the greater Boston and Cambridge area so special? We believe that much of Massachusetts's extraordinary history of innovation can be traced directly to the tensions and interplay among three powerful sectors of American society: academia, the federal government, and industry. Some aspects of the vital roles these three sectors play in Massachusetts are highly visible: Students, professors, and researchers churn out new ideas; the federal government provides the means and the opportunity for many of these ideas to bear fruit; and companies spring up to bring these fruits, or products, to market. The example of Ken Olsen at the beginning of this chapter is an apt illustration of the process. Olsen developed many of his groundbreaking ideas for the minicomputer while working as a researcher at MIT-affiliated Lincoln Laboratory—supported by federal funds—then spun off his ideas to form a new company, Digital Equipment Corporation.

This scenario does not, however, capture the depth or the richness of the functions, roles, and manifestations of each element that allow the trio to interact so productively. Following is a more detailed look at each of these sectors in Massachusetts.

## ACADEMIA

Cambridge has been a college town for more than 350 years. Today, the Boston and Cambridge area has one of the world's densest concentrations of institutions of higher learning. Spearheaded by MIT with its technological prowess, on the one hand, and Harvard with its Ivy League status, and ten world-class graduate schools on the other, the sixty-five colleges and universities in the greater Boston area comprise a resource that lies at the very heart of the area's excellence in innovation.

Although this entire educational complex is crucial to the region, one school in particular, MIT, has been inextricably tied to the creation and growth of the Boston area's new technology community since its founding in 1861. Any examination of the Route 128 region, by necessity, also becomes an examination of the policies and practices at MIT that have made it such a fertile source of ideas, entrepreneurs, and innovative relationships with industry and government.

For example:

• MIT was the nation's leading research organization in the race to mobilize for World War II. The laboratories established there during the 1940s became a springboard for entrepreneurs, products, and entire industries.

• MIT has spawned more spin-off companies than any other single institution in the country. In a series of theses coordinated in the 1960s by Professor Edward Roberts of MIT, researchers identified 129 Boston-area companies launched in the post–World War II period by entrepreneurs associated with just three of MIT's academic departments and four of its laboratories. Moreover, the Bank of Boston recently studied 636 existing businesses in Massachusetts alone founded by MIT alumni.

• With a student body of only about 9,000, MIT has the largest research budget of any university in the nation—about $800 million annually. In 1990, the Institute received $455 million in Department of Defense contracts, more than any other nonprofit organization.

• In 1990, MIT became the first institution ever to be granted more than 100 patents in one year.

Ironically, MIT assumed this leadership role in the high-tech community with little forethought or intent. As a leading school of science and engineering, it inevitably became a source of emerging technologies. But because of its unusually practical orientation, MIT also broke new ground by forming early and lasting links with both government and industry, and by experimenting with how these sectors could and should relate.

But, again, MIT has not acted alone. The roles outlined

below give some sense of the enormous contributions academia has made to the greater Boston area's culture and environment.

## Advanced Training

The primary function of universities is to give students the intellectual underpinnings to contribute as professionals in our society. In Massachusetts, universities have provided the full range of professionals—including scientists, engineers, medical doctors, lawyers, venture capitalists, and managers—required to sustain the high-tech community. This diverse and well-educated labor pool has been critical to the fortunes of the state.

Nearly a quarter of a million students are enrolled in colleges and universities in the greater Boston area, and nearly 427,000 students in all attend institutions of higher education in the state. Largely due to the reputation of its prestigious private schools, Massachusetts has been a magnet for talented, ambitious people from all over the United States and the world. Typically, only about 15 percent of the undergraduate classes at MIT and Harvard come from in state, and more than 30 percent of MIT's graduate students are from overseas. Yet, once they are here, a surprising number of these students decide to stay. Almost a third of MIT's 70,000 alumni have settled in New England to work, mostly in Massachusetts. One recent study found that more than 90 percent of electrical engineers working in Massachusetts received their last degrees from universities in the state.

These graduates are not just carrying newly acquired skills to jobs in industry, they are transferring technology. A company that hires a student who has worked on a project in an emerging field of research typically gets more insight into that research than could ever be gleaned from a technical paper or a conference presentation. Moreover, many graduates of area colleges and universities have gone on to become innovators and entrepreneurs, achieving on-the-job breakthroughs, or launching new companies. These are the people whose creative ideas have sustained Massachusetts's centuries-old reputation as a cradle of innovation.

## Research

A driving force behind the growth of the Route 128 high-tech business community has been the continuous stream of discoveries and insights that have emerged from the area's universities. Research in a broad spectrum of fields of science, engineering, and management has laid the groundwork for commercial enterprises ranging from consulting firms and defense contractors, to computer manufacturers and biotech start-ups. SofTech, for example, sprang from research on numerical control; Repligen from biotechnology; Symbolics from artificial intelligence; Wang from pioneering computer work at Harvard; and American Superconductor from materials research. The founding team at Itek included the entire research staff of a section of Boston University's physics department.

In addition to specific companies, entire industries have arisen from scientific inquiry at universities, ranging from guidance systems, to computers, to biotechnology. In each of these cases, the original research drive was primarily intellectual, not commercial. University researchers were trying to understand the underlying principles governing analog computing, digital computing, or cell behavior, for example. This is not to claim that profit plays no role as a research motivator. The wealth of companies that are offshoots of university work clearly refutes that. But rarely has the question of commercialization or marketability been the foremost concern in plotting the course of a university research project. Precisely this need to remain untainted by short-term rewards has earned universities the reputation for being "ivory towers," isolated from the push and shove of the real world. But the independence and freedom to explore the frontiers of knowledge without a guaranteed payoff has also led to many unanticipated benefits.

## Links to Industry

Although most engineering schools in the Boston area have formed links to industry, it is particularly interesting to look at the experience of MIT, which has been a leader of such interaction for more than a century. Because the Institute was founded on the belief that a practical, real-life orientation would benefit its students, MIT has sought contact with the industrial community throughout most of its history. Early pioneering interactions included programs to place student interns in factories, and regular consultations between faculty members and industry—a practice that became widespread. When Institute President Karl Taylor Compton ruled in the 1930s that professors could spend 20 percent of their time on outside consulting, this was done to rein in their outside activities, not to encourage them.

Today, contacts with industry continue, from the formal connection of the Industrial Liaison Program, which allows member companies to learn about ongoing MIT research, to the informal connection that exists between the Institute and its graduates who have gone on to become entrepreneurs. In addition, MIT has the largest volume of industrially sponsored research on its campus of any university in the world, with the majority of this funding coming from companies located outside the state. Thus, the transfer of ideas both into and out of MIT involves a broad range of companies and a wide geographic spread.

## Links to the Medical Community

A cornerstone of modern medical education is the training provided in teaching hospitals. This integration of education with practice is without comparison in any other field. In the Boston area, one of the largest academic health communities in the nation, the medical schools of Tufts, Boston University, and Harvard have made this practical academic–medical connection especially fruitful. Harvard alone has more than a dozen affiliated teaching hospitals, research centers, and patient care facilities in the area. And MIT, which does not have a medical school, is nevertheless closely tied to these research activities through such collaborations as the Harvard–MIT Division of Health Sciences and Technology Program.

These institutions have gained a reputation not only for providing care and medical education, but also for generating innovative research and medical advances, and even spinning off new companies. Many of the more than forty-five biotechnology firms in the Cambridge and Boston area either had their roots in the local medical establishment or maintain direct ties with the medical research community.

This medical research establishment has won global recognition. The National Institutes of Health sponsors more biomedical research in the Boston area than in any other metropolitan area. In 1989, the state received more than 10 percent of NIH's R&D funding, the highest level of funding in the country on a per capita basis. Moreover, area hospitals are sought after for their industrial research capabilities. Massachusetts General Hospital, for example, the renowned Harvard-affiliated teaching hospital, has accepted a number of industrial research contracts, including an $85-million project for the Japanese pharmaceutical company Shiseido.

## Links to the International Community

Modern university research is an international, cooperative enterprise that involves people and institutions from virtually all nations. Researchers tend to work independently or in small groups, but results and progress are shared in conferences and meetings held all over the world, as well as in an array of scientific publications. As the home of a number of world-class institutions, Boston is thus intimately tied to the international network of scholars, tapping in to the best minds at work in any particular field. This large-scale cooperation is extraordinarily enriching. It serves not only to validate research, but to welcome new ideas, approaches, insights, and ways to interpret the same data. Moreover, it has helped to bring talented people and their ideas into the area.

## THE FEDERAL GOVERNMENT

The federal government, through its various agencies, funds research and product development in a vast array of fields, and Massachusetts consistently ranks among the top five states in federal research dollars awarded. Although the Department of Defense accounts for nearly two-thirds of federal R&D funding in Massachusetts, federally sponsored projects range from the social sciences to virtually every engineering discipline. This research drives the entire process of innovation, underwriting the training of new generations of engineers and scientists. Its value lies not only in its sheer volume, but also in its diversity. The federal government also fuels the Bay State's economy through procurement contracts, both at large corporations, like Raytheon and EG&G, and at a broad network of small subcontractors. Finally, federal policy on matters ranging from the capital gains tax to nationwide research priorities helps determine what kind of environment start-ups will confront.

By focusing on the federal government, we do not mean to suggest that state government has had no role in the growth of the Route 128 region. Only within the last fifteen years, however, has the state attempted to address directly a high-tech community that had been evolving—and largely thriving—on its own for decades. Moreover, recent experience suggests that because state policies are tied to short-term political and economic considerations, a program touted as essential to economic growth one year may be abandoned the next. Even the state's support for higher education, acknowledged by most observers as critical to the Commonwealth's high-tech competitiveness, has been severely scaled back during the recent downturn in the state's economic fortunes.

## University Research Contracts

In the United States, the federal government funds a vast proportion of basic research, most of which is conducted in universities. Despite concerns about future funding patterns, the government's past support of university research has already created the most extensive basic research infrastructure in the world. Federal funding is largely responsible for making the American university system the envy of other nations. With its collection of top-notch research institutions, Massachusetts is a crucial node in this system.

A number of different federal agencies, with vastly different interests and missions, sponsor university research, including the National Science Foundation, the National Institutes of Health, the National Aeronautics and Space Administration, the Federal Aviation Administration, and the Departments of Defense, Transportation, and Energy. Since World War II, and particularly since the founding of the National Science Foundation, the federal government has sponsored this research in the anticipation of the long-term benefits it may hold for society.

Broadly, "basic" research consists of scientific investigations that seek to expand our fundamental knowledge and understanding of ourselves and the world around us. A consequence—and often a motivation—of this understanding is the ability to alter or control some process or phenomenon. Ultimately, when we apply this ability toward some useful end, a technology is born. Basic scientific research is often, therefore, at the very root of technological innovation.

Yet the outcome of scientific research is inherently difficult to predict; many scientific advances have occurred by accident rather than by design. A particular line of research may uncover entirely unanticipated benefits, or, conversely, may fail to produce any useful results at all. This highly haphazard process relies on exploring as many avenues as possible in order to uncover new knowledge and insights that may find useful applications, but the results can be impressive. Boston-area researchers backed by federal funds have been behind the emergence of such key new fields as biotechnology, artificial intelligence, and advanced materials, such as polymers, ceramics, and metals with new and more predictable properties. The results of this research have been transferred to industry and government—both in the region and beyond—often providing the impetus for start-up companies.

Federal research—in typically American fashion—is awarded on a competitive basis: University researchers must submit their ideas and proposed research plans to appropriate agencies for peer evaluation and approval. Although some criticize this system as too political and time-consuming, it compels researchers to articulate the importance of their work, and, in a very real sense, forces them to be entrepreneurs.

The specific direction of university research is determined primarily by the principal investigator, not by the funding agency. Although the federal government may set broad agendas within the context of energy, transportation, or defense, for example, for the most part researchers are allowed considerable latitude in what specific scientific and technical questions to address.

## Federal Laboratories

In addition to sponsoring university research, the federal government has created a network of laboratories targeted to its specific needs, including the development of specialized defense systems and other classified work, and Massachusetts has been a favored venue for some of this research. By siting labs near the Cambridge/Boston university complex, the government is continuing the cross-fertilization of government and academic interests that was so productive during World War II. In fact, MIT runs Lincoln Laboratory, the region's largest federal lab with an annual research budget of about $400 million, for the Air Force. Although the actual links between Lincoln Laboratory and the MIT campus have decreased since the lab's founding in the early 1950s, both organizations have strong and complementary roles in the region.

The extended research community provided by federal laboratories creates both a market and a training ground for young scientists and engineers, allowing them to hone their skills in advanced technologies by solving challenging problems. In addition, like universities, federal labs cultivate new ideas that often create opportunities for entrepreneurship: Some fifty companies spun off from Lincoln Laboratory in just over a decade.

## Other Research Contracts

In addition to basic research at universities, and the more goal-oriented research at federal laboratories, the government also funds research at a variety of independent organizations in the Route 128 region, such as the nonprofit MITRE and Draper Laboratories, and a number of corporate laboratories. Once again, these labs provide a springboard for entrepreneurship. Not uncommonly, engineers and scientists working in this extended infrastructure have felt they could deliver the same research services more competitively on their own. Alternatively, they may have seen a wider market for their expertise, or may have felt that a particular project that has lost federal funding deserves to be pursued. In any case, the result is new companies spinning off from the research network.

## Procurement Contracts

Though sometimes downplayed by state politicians, federal contracts—in particular from the Department of Defense—have been an important component of the regional economy for decades. Defense contractor Raytheon, and EG&G, a diversified technology company that works closely with the Department of Energy and the National Aeronautics and Space Administration (NASA), are just two of the major companies in the state that rely on government contracts for a large part of their sales and earnings. Massachusetts organizations received more than $8 billion in DOD prime contracts in 1990.

In addition to these highly visible corporations there is a network of hundreds of small specialty subcontractors that may do little or no commercial work, relying solely on government business. Indeed, many start-ups have arisen specifically to serve small sectors of the government market.

## Policy

At the highest level, the federal government creates policies that affect the overall environment in which innovation takes place. For example, the government sets trends in research by determining the broad context of where funding will be most plentiful. In the 1970s, federal support for alternative energy schemes encouraged research in such areas as photovoltaics and synthetic fuels. The following decade, the Strategic Defense Initiative—or Star Wars program—supported hundreds of millions of dollars of research in Massachusetts. The technological requirements of such a sweeping program sent area research organizations scrambling to tailor their areas of expertise to meet the government's interests.

The federal government has also instituted a number of specific programs and policies aimed at encouraging and supporting entrepreneurship. These include the Small Business Investment Company program, launched in 1958 to help get capital to small businesses; the Small Business Innovation Research program, introduced in 1977 to encourage federal agencies to award research contracts to small businesses; and tax policies, such as cuts in the capital gains tax, aimed at encouraging start-up investments.

These federal activities have been useful, but it is important to note that the greater Boston area has weathered a number of changes in both research trends and policy directions. The government's primary role in affecting the flow of innovation has been in the volume and diversity of its research.

## INDUSTRY

There are more than 3,000 high-technology firms in Massachusetts. These businesses, in concert with the academic complex and the government-supported infrastructure, form the critical mass that has made the Route 128 region a mecca of innovation.

### Start-ups and Small Firms

New entrepreneurial high-tech firms in Massachusetts garnered a great deal of attention during the 1980s. These aggressive start-ups were seen by many as holding the promise of new jobs, new markets, and continued regional prosperity. The steady growth of a company like Digital Equipment from its humble beginnings in the 1950s and the rapid-fire success of Lotus Development over a two-year period seemed to augur well for high tech's ability to support the regional economy. These success stories helped fuel the reputation of the Massachusetts Miracle.

These companies are sunflowers in a field of buttercups, however. Few enterprises achieve either the long-term success of a Digital or the explosive ascent of a Lotus. Most companies in Massachusetts and elsewhere start small and stay small, if they survive at all. More than 70 percent of software firms in the state employ fewer than twenty-five people, according to the latest survey by Mass Tech Times, publisher of the *Mass High Tech* newspaper, and more than 60 percent of manufacturing companies employ fewer than fifty. Although this broad network of small companies has a significant cumulative impact, most of these firms will never grow large enough to have a direct bearing on the regional economy.

The real value of these firms lies in their willingness and ability to take new technologies—in the form of an idea, an approach, or just expertise—to the marketplace. MIT's Technology Licensing Office, for example, reports that 85 percent of MIT patents are licensed to small firms. In addition, entrepreneurship can thrive even during hard times. The tight venture capital market of the 1970s did not stop the founders of Prime Computer and Kurzweil Computer Products, for example. And when NASA's space program wound down in the late 1960s, a number of high-tech companies were formed either in anticipation of layoffs or because of them. Would-be entrepreneurs who have been thinking of testing the waters are sometimes tossed in, where they must sink or swim.

## High-Tech Giants

Companies that have "made it"—grown large and successful—are the highly visible mainstays of the region's high-tech establishment. Digital Equipment, Raytheon, EG&G, Thermo Electron, and Lotus Development are among the state's largest employers and play a significant role in the economic well-being of the region. When their businesses are thriving, a positive ripple effect extends to the community of service and support firms that surrounds them—ranging from subcontractors to law firms to restaurants. When their profits decline, the network of support companies suffers, too.

The significance of these high-tech giants goes well beyond the economic impact of the people they employ and the services they buy, however. Mitchell Kapor, founder of Lotus, has likened the region's minicomputer makers to "factories that develop people."[2] Other high-tech firms looking to hire in the region have a built-in, highly trained pool of labor to draw from.

Companies like Digital, EG&G, and Lotus—along with universities and government labs—are a key source of entrepreneurs for the region's start-ups. Hundreds of engineers, software developers, and managers have turned their backs on established businesses, convinced that they could "do it better" themselves. Through this process Digital Equipment begat Data General, which begat Apollo Computer, which, at recent count, had parented nine other high-tech firms.

## Service and Support

The infrastructure of service and support companies that has gradually grown up around the Boston area's high-tech community is one of the key ingredients that differentiates the area's business environment from that of other regions of the country. The intense concentration of venture capital firms, for example, has been both a result of and a catalyst behind Route 128's development.

Other, less visible businesses have also paved the way for the rapid expansion of the high-tech community, however. Machine shops, banks, and law firms all understand the needs and expectations of high-tech firms and have shaped their offerings to meet them. Perhaps even more important, this support network accepts the high-risk nature of many new technology firms.

## MAKING IT WORK

Some people have tried to characterize the connection between the three sectors we have just described—academia, government, and industry—as a partnership. In fact, the relationship of the three over the years has been more a haphazard process of coming together to address projects or issues of mutual interest, and then breaking apart again in response to the tensions and pulls of their fundamentally different missions. Although not always harmonious or predictable, this relationship has been highly productive. The coming together and breaking apart of these three forces has been a key catalyst behind innovation—spinning off ideas, people, products, and companies. Moreover, the contributions from this fertile union visible today are just the tip of the iceberg. Government, academia, and industry have a tradition of interactions in the state that, because of Massachusetts's long and rich history, is unduplicated.

In examining these interactions, we have chosen to begin with the Morrill Land Grant Act of 1862, a pivotal piece of federal legislation that for the first time directly linked the fortunes of the government, universities, and industry. The Act's passage coincided with the founding of MIT one year earlier. Together, these two events set the stage for the innovation that has been occurring in Massachusetts ever since.

No one can doubt that this creative process—the generation of new technologies, their application in industry, and the spawning of new companies—has become a vital component of a competitive, healthy economy. In the last few decades, advances in computers and software have transformed not only the way the world does business, but also the way we live.

Biotechnology and new communications technologies are now delivering on their early promise. And such emerging technologies as high temperature superconductivity and solar power show even more promise for the future. In all these fields, the Route 128 region has excelled, and it is the force of this excellence, coupled with the practical drive to put ideas to work, that in the 1980s helped carry the region's economy forward in its wake.

Route 128 is more than just a stretch of highway dotted with companies, though. It is a creative phenomenon born of—and sustained by—the complex interactions of government, academia, and industry. Only by examining the people, forces, and events that shaped Massachusetts can we come to understand the true nature of this resource. And understanding it is critical. Unless we know what attitudes, actions, and policies led to its development we can inadvertently weaken or even destroy it. On the other hand, if we learn its secrets, we can make the phenomenon stronger, more productive, and more widespread. Moreover, the lessons we draw from this experience can help the United States in its current struggle to nurture its strength in innovation, to become more competitive, and to make the most of its existing resources.

The story of eastern Massachusetts's high-tech community is a story of idealism and entrepreneurship, of ivory tower intellectualism and practical Yankee ingenuity, of cultures and nations in conflict, and of individual dreams and cooperative efforts. It is a story of the search for new knowledge and the drive to put it to work.

# 2

# Creative Tension: Industry, Academia, and Government

IN 1862, President Abraham Lincoln had a lot on his mind. A year earlier, Fort Sumter had fallen to hostile Southern troops, pitching the nation into the Civil War. On July 1, 1862, President Lincoln issued a draft call for an additional 300,000 Union troops to bolster his flagging army. Yet, the following day, the President found the time to sign the Morrill Act, setting in motion a plan that spurred the creation of a network of colleges and universities, and encouraged the development of new curricula that would help make the United States the most productive agricultural nation in the world. Moreover, the Act set a precedent as the first formal collaboration among industry, academia, and government—both federal and local.

The federal government held vast tracts of land throughout the nation; under the Act, the government gave nearly 17.5 million acres—16,000 square miles—to the states, which were to use revenues from the land to subsidize the development of a system of state colleges with curricula aimed at supporting the needs of local industry.

Specifically, states were to support and maintain

> at least one college where the leading object shall be, without excluding other scientific and classical studies and including military tactics, to teach such branches of learning as are related to agriculture and the mechanic arts, in such a manner as the legislature of the States may respectively prescribe, in order to promote the liberal and practical education of the industrial classes in the several pursuits and professions in life.[1]

Significantly, the government never had much influence on the use of the funds. Instead, the states were allowed the latitude to apply the monies toward the Act's broadly defined goal. In effect, the Morrill Land Grant Act subsidized the development of curricula directly related to the broader needs of society, and, in particular, of industry. Academic institutions, once focused exclusively on the classics and religion, evolved a vast array of courses and degrees in the sciences, engineering, and management. Nowhere was this more true than in Massachusetts.

## THE MORRILL LAND GRANT ACT

Education was well established as a high priority among mid-nineteenth-century Americans, particularly in New England. From the start, Boston's colonists—many of whom were graduates of Cambridge or Oxford—cared deeply about education. As the author of a pamphlet titled "New England's First Fruits" put it in 1643, "After God had carried us safe to New England, and we had builded our houses, provided necessaries for our livelihood, reared convenient places for God's worship, and settled the civil government: one of the next things we longed for, and looked after was to advance learning and perpetuate it to posterity."[2] In 1636, just sixteen years after the Pilgrims disembarked at Plymouth Rock, the general court of Massachusetts founded Harvard College, the colonies' first institution of higher learning.

Although the parochial Harvard of its first two centuries bore little resemblance to the modern university it began to evolve into in the early 1900s, the college came to symbolize the early New Englanders' determination to nourish the minds of the region's inhabitants. By the 1860s, Boston had already been dubbed the "Athens of America." Harvard College had been joined by numerous other institutes of higher learning in New England, including Boston University, Tufts, Amherst College, and Williams College, and Boston had become renowned for such writers and thinkers as Ralph Waldo Emerson, Bronson Alcott, Louisa May Alcott, Nathaniel Hawthorne, and Margaret Fuller.

The sheer number of colleges that sprang up in the region was a testament to the importance early Americans attached to both religion and education. A shortage of ministers in the colonies was an early driver of higher education. Harvard was not formally affiliated with a specific religious denomination, but the desire to produce graduates schooled in specific religions contributed to the density of colleges in the greater Boston area. Boston University, for example, was founded in 1839 as the first college in the United States to train Methodist ministers, Tufts University was founded by members of the Universalist Church in 1852, and both the College of the Holy Cross in Worcester and Boston College were founded as Jesuit colleges of higher education. As recently as 1948, Brandeis University in Waltham was founded by the American Jewish Community, although it is operated as a nonsectarian school.

Settlers carried this tradition across the nation. In addition to older, established Eastern and Southern schools, several hundred smaller colleges sprang up in small towns across the Midwest. These schools, many of which did not ultimately survive, reflected the prevailing view that broad access to higher learning was an important right of the citizenry of a region, and perhaps the key to establishing an area's prestige and image as a bona fide outpost of civilization.

One believer in education for the masses was Jonathan Baldwin Turner, a Yale graduate, who joined the faculty of a tiny college in Illinois in 1833. Turner's training was in Greek and Latin, and he was hired to teach rhetoric, but he did not forget his early roots as the son of a Massachusetts farmer. Recognizing the need for an efficient way for farmers to fence in their land on the treeless prairie, he experimented with various species of plants until he found one that could make a fast-growing, impenetrable hedge. Turner was eventually fired from his job at the college for his antislavery views, but his idea of a living fence fostered a flourishing business.

The success of this venture in meeting the needs of the community convinced Turner that universities could—and should—play a role in educating farmers and other workers, not just doctors, lawyers, and other professionals. He became an ardent advocate for tax-supported state schools and the democratization of education. Inspired by Turner's persistence, the Illinois state legislature petitioned Congress in 1853 for federal lands to help set up a university.

Turner's ideas struck a chord with Justin S. Morrill, a Republican congressman from Vermont who also saw the political value of using the issue to build the new Republican party's strength among the industrial classes. He drafted the bill that Abraham Lincoln eventually signed into law. Although it may seem odd that such pivotal legislation should have been passed in the midst of the Civil War, in fact, the Morrill Act was only one example of broad changes enacted by the second session of the 37th Congress, which historian James M. McPherson claims, "did more than any other in history to change the course of national life."[3] In the same year that the Morrill Act was passed, Congress also passed the Homestead Act—granting public land to settlers, and thereby fueling westward expansion—and the Pacific Railroad Act—providing land and government loans to complete a transcontinental railroad. By these acts, the federal government conclusively showed its power not only as a source of capital, but also as an agent for economic and social development.

## THE ACT TAKES HOLD

The land-grant concept of the Morrill Act breathed new life into the network of colleges and universities in America, cutting back drastically on the mortality rate of the tiny institutions scattered across the land. Most funds from the program initially helped universities that were either already in operation or just starting up at the time, including Brown, Rutgers, and Yale as well as state colleges in Georgia, Delaware, Iowa, Michigan, Missouri, and Wisconsin. By 1880, land grants had helped start state colleges in eleven states, eight new agricultural and mechanical colleges, and the now-prestigious private school, Cornell University. Attendance at land-grant colleges was 2,243 in 1882, but it had grown to more than ten times that, almost 25,000, by 1895. Enrollment continued to soar in the twentieth century, hitting 400,000 in 1945, and topping 1 million by 1988. A second Morrill Act was passed in 1890 that established annual federal appropriations for land-grant colleges. State legislatures, however, seized on the political benefits of supporting the schools, increasing their funding at a faster rate. By 1932, the federal share dropped to one-tenth of the total, down from one-third in 1910.

The philosophy behind the Morrill Act struck a chord with an intellectual trend emerging at the time. A number of debates had begun on the potential value of approaching "the practical arts"—as opposed to the fine arts—in a more systematic manner. In those days, the practical arts meant the broad range of useful activities that included such crafts as toolmaking, tinsmithing, weaving, and the like. Some of these crafts—especially textiles—were beginning to make the transition to more formal manufacturing, and many thought that a more organized approach to the underlying body of knowledge, including the application of scientific principles, would lead to significant improvements in the techniques and processes involved. In an 1829 publication, Harvard physician Jacob Bigelow used the obscure word "technology"—derived from Greek words meaning "craft" or "systematic treatment"—to label this higher level approach to practical endeavors, and the word eased its way into more general usage.[4]

The application of systematic approaches to practical problems was far more than just an intellectual trend, however. Knowledge of mathematics, mechanics, and materials had long been an integral part of the practice of civil engineering, for example. And the chemical industry was increasingly coming to rely on an improved understanding of the underlying chemistry involved in the production of its products, or the improvement of their performance.

In the 1830s, the striking properties of rubber had fueled a boom in companies making products ranging from flexible pipes to waterproof coats. One of the pioneers in this new industry was the Roxbury India Rubber Company, set up just south of Boston. The failure of rubber products to live up to their promise because of the instability of the substance at both low and high temperatures, coupled with an economic downturn in 1837, drove most companies out of the business. In 1839, however, in a remarkable feat of persistent empirical chemistry in his factory in Woburn, Massachusetts, New Englander Charles Goodyear stumbled on a way to stabilize rubber. But even after he knew the basic approach, it took months to develop a process that produced a consistent product. Further refinements and enhancements to meet the needs of specific applications would take years of research.

The merger of academic research and learning with practical pursuits proved a potent combination. Out of the agricultural land-grant colleges arose better planting techniques, new crop strains, improved livestock management methods—in sum, a far more systematic, or scientific, approach to agricultural production. The Morrill Act also gave a crucial boost to the incorporation of science into university curricula, and to the development of diverse emerging fields of engineering. After all, engineering is a bridge between theory and practice. There were only six engineering schools in the United States when Congress passed the Morrill Act, but twenty years later there were eighty-five, half of them the engineering departments at land-grant colleges. On the eve of World War I, there were 126 university-level engineering schools turning out more than 4,300 engineers a year, up from just 100 graduates in 1870.

Once the idea of adapting colleges to include more practical learning became well established, the greater Boston area blossomed with a variety of new institutes of higher learning aimed at specific vocations or populations. For example, the Lowell Technological Institute, a precursor to the University of Lowell, responding to the needs of that mill town, was founded in 1895 as a proprietary textile school, devoted to the arts and sciences of weaving, dyeing, engineering, and chemistry in textile production.

Northeastern University, founded in 1898, grew out of an "Evening Institute for Young Men" sponsored by the Young Men's Christian Association (YMCA). In 1909, the school adopted the same concept of cooperative education that distinguishes it today, in which students alternate traditional classroom studies with related work experience. Simmons College, incorporated in 1899, was not only an early women's college, but one of the first to focus on the practical goal of helping women make an independent livelihood. Wentworth Institute of Technology, founded in 1904 as a two-year mechanical arts school, promised young men quick access to a job in the area's burgeoning industrial community.

The proliferation of such schools in eastern Massachusetts was perhaps inevitable, given the region's well-established academic reputation and its role at the time as the capital of American industry. One of the best examples of a school that embraced the practical orientation espoused by the Morrill Act was the Massachusetts Institute of Technology (MIT), founded in 1861, a year before the Act was signed.

## INDUSTRY IN NEW ENGLAND

As is clear from its wording, the Morrill Act's intent was to improve the learning of all "industrial classes." But because local industry in most of America meant farming, the Act had a particularly strong impact on the study and practice of agriculture. The petroleum industry didn't exist until the discovery of oil in Pennsylvania in 1859, steel was in its infancy, and automobiles hadn't been invented yet. Industry and finance were concentrated in New England, specifically Boston, which from 1850 to 1900 grew from a seaport of about 200,000 people to a sprawling industrial metropolis of more than a million. Its flourishing industries included shipbuilding, textiles, and machine tools.

From the start, the culture and economy of New England were molded by the particular limitations—and strengths—of its resources. The region's rocky soil and harsh winters dictated against an agriculturally based economy. Instead, while regions to the south increasingly became defined by a particular crop, such as tobacco or cotton, New England was defined by the sea. Fishing—for bass, lobsters, haddock, crabs, mackerel, and, particularly, for cod—dominated the region's economy up to the Revolutionary War. Then followed the fortune-building years of the China trade. With their increasingly sophisticated ships, traders set off for China and many other parts of the world. Because New England had no natural products of its own to export, resourcefulness was the key. Some traders ferried products from one foreign port to another. Despite the region's later fame as a center for abolitionists, slaves were a key product in the so-called triangular trade in which Massachusetts-made rum was traded for African slaves, who were then traded in the West Indies for the raw materials necessary to make more rum. Others brought cargoes of cocoa, spices, or jute back to Boston, processed them into finished products, and shipped them off again.

This early reliance on the sea created not only a particular way of life, but also a sort of regional receptivity to risk taking, exploring new ideas, and finding opportunities in unlikely situations—in other words, "Yankee ingenuity." Those traders who prospered were known to be quick-witted, decisive, and deft at spotting a need and finding a way to fill it.

Famous examples of resourcefulness and entrepreneurship also existed among those who stayed on land. Frederick Tudor of Boston, later known as the "Ice King," decided to act on his brother's playful suggestion in 1811 that New England pond ice should be harvested and sold in the Carribbean. Tudor lost nearly half of the $10,000 he invested in his first venture, to ship 130 tons of ice to the tropical island of Martinique, and it was fifteen years before his ice trade was profitable. But by 1856, New England ice had become a worldwide commodity, with more than 130,000 tons being shipped from Boston to other parts of the United States and abroad.

Tudor's business achievement was a feat of both technical and marketing acumen: To make his venture a success, he not only had to design efficient icehouses to prevent melting in tropical climes, and to engineer a machine to cut pond ice into uniform and easily harvested chunks, he also had to persuade people in every port his ships could reach that ice was something they wanted and needed.

The very stone underfoot that made the land so inhospitable to farming also became an export product. At the end of the eighteenth century, an improved technique for cutting granite—pioneered by a workman near Salem—transformed this stone into a practical building material for the first time. As New Englanders wrestled with and perfected better ways to cut and handle the stone, granite became coveted, particularly in the Eastern and Southern United States, for the solid and classic feel it gave to houses, public buildings, and churches. The demand for granite—both in the form of large blocks and smaller paving stones—prompted Charles Francis

Adams, Jr., author and prominent businessman, to declare in 1883 that, "the three great staples of New England are ice and rocks and men."[5]

But it was the birth of the American textile industry and the creation of the first modern factory that established New England's industrial leadership, and helped the Boston area flourish as a cradle of innovation. When Francis Cabot Lowell, a Boston merchant, made an extended visit to England in 1811, his imagination was captured by the possibilities of that country's water-driven power looms for weaving cloth. His friend and future business partner, Nathan Appleton, recounted that after meeting Lowell in Edinburgh, "[Lowell] informed me that he had determined, before his return to America, to visit Manchester, for the purpose of obtaining all possible information on the subject, with a view to the introduction of the improved manufacture in the United States."[6] The design of the looms was zealously guarded, and Lowell carried no physical plans with him back to Boston.

Nevertheless, drawing on his remarkable memory, and working with an expert mechanic named Paul Moody, Lowell re-created the power loom he had studied overseas. Even more impressive is what he did with it. The looms Lowell studied in England were isolated, a discrete step in a sharply defined production process. Lowell abandoned that structure, uniting all the main steps in textile production—from carding the raw cotton to weaving the finished cloth—under one roof. It was a deceptively simple, but profound organizational innovation.

Ironically, the notion of bringing these functions together had already been conceived in Britain, but the established structure of the textile industry prohibited its implementation. This was not the case in America. Lowell's textile factory, established just west of Boston in Waltham in 1814, an area subsequently reenergized by Route 128, created a model for the modern factory.

Lowell's enterprise soon moved beyond Waltham. Because the Charles River, which meets the sea at Boston Harbor, supplied only enough power for three mills, Nathan Appleton and Lowell's other partners, a group that came to be known as the Boston Associates, began scouting for new sites in 1821. Appleton and his cohorts settled on a stretch of the Merrimack River, running through the farmlands of Chelmsford, that dropped thirty feet in less than a mile. The first wheel of the Merrimack Company began turning in September 1823, the first cotton goods were shipped to Boston five months later, and the town of Lowell was incorporated around the mills in 1826. By 1840, the factories at Lowell were among the most productive in the world.

The "American style" of manufacturing pioneered in Lowell's mill was soon employed to produce other products such as guns, machine tools, boots, and shoes. Immigrants poured in, industry flourished, and during the second half of the century, Boston reveled in its self-proclaimed status as "Hub of the Universe."

This concentration of industry, coupled with the intellectual vitality of the region, attracted educator William Barton Rogers to Boston in 1853. He and his three scientist brothers were strong advocates of the principles of technology, the blending of science and the useful arts. Since 1828, they had often talked about a new kind of technical school, one that would serve the needs and opportunities created by the Industrial Revolution by giving scientists, engineers, and other technical workers a scientifically sound but practical education.

Rogers's own background was varied and eclectic, and, in a sense, mirrored the marriage of science and industry he hoped to achieve in his school. Before moving to Boston, Rogers had taught natural philosophy (science) and chemistry at the College of William and Mary; had served as advocate and chief of the first Geological Survey in Virginia, doing much of the legwork himself; had lectured extensively on a variety of scientific topics both in the United States and in Europe; and had been elected chairman of the faculty at the University of Virginia, where he was a professor of natural philosophy.

As his reputation as a scientist and educator grew, Rogers remained captivated by the idea that America needed an institute devoted to technological studies—something on the scale of the grand European museums, technical societies, and colleges of technology that he had visited when lecturing overseas. Although America was approaching the forefront of the Industrial Revolution, the nation's schools of higher education had not kept pace. Indeed, by 1846, there were only half a dozen technical schools, including West Point and the predecessor to Rensselaer Polytechnic Institute, that were granting degrees in military or civil engineering.

In 1846, Rogers's brother Henry, then living in Boston, wrote that the opportunity seemed to have come. The Lowell Institute, established ten years earlier by the will of John Lowell, Jr., a Boston merchant and son of Francis Cabot Lowell, was already sponsoring free public lectures, and was considering adding a technical school. Henry urged his older brother to write down his ideas for such a school to be presented to John A. Lowell, trustee of the Institute.

Rogers's response, titled "A Plan for a Polytechnic Institution in Boston," set forth his philosophy:

> We may safely affirm that there is no branch of practical industry, whether in the arts of construction, manufactures or

> agriculture, which is not capable of being better practised, and even of being improved in its processes, through the knowledge of its connections with physical truth and laws, and therefore we would add that there is no class of operatives to whom the teaching of science may not become of direct and substantial utility and material usefulness. . . .
>
> A polytechnic school, therefore, duly organized, has in view an object of the utmost practical value, and one which in such a community as that of Boston could not fail of being realized in the amplest degree.[7]

The plan for the Lowell Institute went no further than this proposal, due to restrictions in Lowell's will. But other efforts were afoot in the region as well. The following year, with a $50,000 grant from textile magnate Abbott Lawrence, Harvard launched the Lawrence Scientific School to apply the principles of science to engineering, mining, and machinery design, and Yale opened a science department that would later become its Sheffield Scientific School. As might be expected, however, these "practical" initiatives were not embraced wholeheartedly by their respective academic communities, and pragmatic curricula were slow to develop. Although academia felt comfortable with scientific principles, it preferred to keep this more nuts-and-bolts training at a distance. Forty-five years later, only 150 engineers had graduated from Harvard.

In any case, Rogers believed that these ideas didn't go far enough. He saw a need for an independent institute at which students would be taught how to apply scientific principles to industrial problems. In 1859, as new land was created in Boston by filling in the shallow basin of the Charles River known as the Back Bay, Massachusetts's governor announced that part of this land could be used for "such public educational improvements as will keep the name of the Commonwealth forever green in the memory of her children."[8]

Two broad plans—backed by Rogers—to establish educational services in the Back Bay failed to win legislative support. But a third and more limited plan presented by Rogers for an "Industrial Institution designed for the advancement of the industrial arts and sciences and practical education in the Commonwealth" succeeded. Rogers's plan, based on his proposal thirteen years earlier to the Lowell Institute, and backed by educational, business, and professional leaders of the Boston community, was granted a charter, and on April 10, 1861, the governor approved the "Act to Incorporate the Massachusetts Institute of Technology."

On February 20, 1865, fifteen students began classes at the new Massachusetts Institute of Technology located in rented rooms in downtown Boston. The total initial enrollment reached twenty-three, and the first class to graduate numbered thirteen. Among the early students were John M. Forbes of the famous shipping family, and Joseph W. Revere, grandson of Paul Revere. By the time its doors opened, the school had gotten a critical boost from the Morrill Act. Although most of the state's allotment went to support the Massachusetts Agricultural College, founded in 1863 in Amherst in western Massachusetts, a much-needed three-tenths of the funds ended up at MIT.

In a departure from the practice at most institutions of higher learning at the time, classes at MIT blended textbook studies with experience in the laboratory as well as visits to machine shops, engines, mills, furnaces, and chemical works. Rogers and his colleagues believed that this strong practical orientation better prepared graduates for the challenges they would encounter as working professionals in any field. This initial closeness between theory and practice and this sensitivity to the needs of industry have been key influences on the evolving character of the Institute ever since.

Three years after it opened its doors, the Institute was considered a success. The opening of MIT marked the fusion of the industrial and academic traditions in New England under the auspices of the federal government through the Morrill Act. This fertile union set the stage for the prolific century of innovation that was to follow.

## THE BIRTH OF INDUSTRIAL RESEARCH

All this activity took place against a backdrop of profound changes in industry, and in the nature of the emerging profession of engineering. Until the nineteenth century, there had been little need for formal scientific training in industry, which tended to be decidedly "low tech." But in the 1800s, the increasing sophistication of problems facing many industries awakened a need for employees with formal training in the emerging techniques of science and engineering. It was these forces that helped win approval of the Morrill Act, that, in turn, helped to accelerate the changes in industry.

Engineering as a profession first emerged from industry, not academia. People who called themselves engineers were specialists in building bridges, roads, and canals (civil engineers), experts in mining operations (mining engineers), or experienced in machine shops (mechanical engineers). The first academic engineering programs were in these established fields. West Point in 1802, and Rensselaer in 1824, both started early civil engineering programs, and Yale in 1846, followed six years later by the University of Michigan, offered the first programs in mechanical engineering, all created in response to industrial need.

Electrical engineering and chemical engineering, on the other hand, sprang from new businesses with less history. Both professions emerged with the sudden growth of the chemical and electrical industries at the turn of the century, industries that relied crucially on chemistry and physics, respectively, to conceive, develop, and refine their products. These businesses were the first to engage in formal research programs.

"Research serves to make building-stones out of stumbling-blocks," once declared Arthur D. Little, the founder of the world's first consulting firm.[9] This sentiment may have been admirably expressed, but when Little first opened for business in Boston in 1886, industrial research was not a tool most companies even considered.

As the status and focus of technical universities developed throughout the 1800s, a few companies appealed to professors for help with research. For the most part, however, industrial research was an unrecognized function carried on by self-motivated—and isolated—individuals. Moreover, formal scientific disciplines such as chemistry were still new enough in the late 1800s that the general public viewed them with considerable speculation. However, largely due to the efforts of a few New Englanders, industrial research began to take hold around the turn of the century as a respected, and critical, corporate function.

Arthur D. Little was central in proving the worth of industrial research. Raised in Portland, Maine, he had already demonstrated his scientific curiosity as a child, winning second prize in a Harvard University–sponsored contest challenging schoolchildren to write essays about plant and animal life along the Atlantic coast. He attended MIT, but had not graduated when he and a partner opened Griffin & Little, Chemical Engineers, on downtown Boston's Milk Street in 1886. An ad that the partners placed in the *New England Grocer* that first fall nicely summarized the scope of their business: "Examinations of Chemicals, Drugs, Paints, Oil, Grease, Soap-stock, Soaps, Fertilizers and Fertilizing Materials, Minerals, Iron and Steel, Water Analysis, Analysis of Foods and Foods Products, and all kinds of general analyses made in the SHORTEST POSSIBLE TIME and at the lowest price compatible with Accurate and Reliable work."[10]

This first consulting firm faced multiple hurdles, not the least of which were distrust and apathy on the part of the local business community. "Industry and science were comparative strangers," Little reminisced almost fifty years later, "and industry had little desire for closer acquaintance."[11] Little made it his life goal to turn that acquaintanceship into a lasting partnership. Thanks in part to his efforts, the chemical industry was the first to embrace industrial research, accounting for more than 25 percent of all new industrial laboratory discoveries in the United States between 1899 and 1946.

Although in the early years Little struggled to gain credibility and to convince the chemistry profession of the importance of industrial research, the message had begun to be heard by the time his firm adopted the name Arthur D. Little, Incorporated (ADL) in 1909. ADL would eventually become the consultant for such corporate giants as United Shoe Machinery, the General Chemical Corporation, and International Paper. In 1911, ADL even established General Motors's first materials testing laboratory, complete with a research staff of seven.

Little's professional life was to keep him in close and continual contact with MIT, though he never completed his degree. When MIT moved to Cambridge in 1916, Little followed suit and built ADL's headquarters right down the street on a stretch that would later earn the sobriquet of Research Row. In 1977 the Department of the Interior declared the building a National Historic Site, honoring its importance as the birthplace of industrial research.

Just as Little helped create legitimacy for industrial consulting, Elihu Thomson and General Electric pioneered the concept of industrial research aimed, at least in part, at scientific discovery. Before the age of industrial research, major chemical and electrical companies typically expanded their technological capabilities—often quashing competition in the process—by buying up patents from independent inventors, or acquiring smaller companies that controlled a process they needed. General Electric, in particular, was infamous for buying what it needed to stay ahead and maintain control of the lighting market. Such aggressive mergers and acquisitions considerably lessened the need for internal technological research. Holes began to appear in this strategy, however, by the end of the nineteenth century as scientists, in both Germany and the United States, continued to make advances in lighting technology that threatened GE's hold on the market.

GE responded to this new reality in 1900 by establishing the first wide-ranging industrial research laboratory. There were already at least thirty-nine industrial labs in the United States by this time, but GE's was the first to undertake fundamental scientific research. To head up this innovative venture, GE recruited Willis R. Whitney, a graduate of both MIT and the University of Leipzig in Germany, who was then a chemistry instructor at MIT. At first, Whitney continued to teach at MIT, but the lab's growth convinced him to join GE full time after three years. What began modestly enough as a handful of people in a carriage house on the banks of the Erie Canal in Schenectady, New York, was soon a thriving concern. By 1906 the lab had 102 people on staff, and by 1929 there were 555. Whitney's crew laid the groundwork for many important GE developments—including the company's move into the radio business—and two scientists from the lab would eventually win the Nobel Prize.

During the 1920s, public recognition of—and appreciation for—science and research may have reached an all-time high. Science was not an abstract concept, but a means to concrete products, such as refrigerators, radios, and other innovations that were making life for Americans better and easier. There were fewer than 100 U.S. industrial research labs before World War I, but by 1929 there were more than 1,000.

Not all these labs followed GE's example of incorporating basic science into applied research. A notable exception was Du Pont. This chemical company had engaged in practical research of one kind or another since 1888, and when three MIT alumni, Pierre, Coleman, and Alfred Du Pont, bought the company from relatives in 1902, they vowed to beef up its research work in order to expand the company's horizons. Despite this pledge, in the first few decades of the twentieth century, Du Pont, like the early GE, was more likely to expand its technological prowess through acquisition rather than by internal development. What research the company did was directed at solving concrete problems.

In 1926, however, the director of Du Pont's chemical department, Charles Stine, submitted a memo to the company's executive committee entitled "Pure Science Work." Citing GE and certain German companies as precedents, and reacting no doubt to antitrust concerns that were making it more problematic to acquire new technological developments from the outside, Stine pressed for the company to undertake research with "the object of establishing or discovering new scientific facts."[12]

The executive committee concurred, committing the princely sum of $25,000 a month to the lab's operations on top of the $115,000 allotted to build "Purity Hall," as Du Pont chemists soon christened it. Although Stine at first found it difficult to recruit the scientists he needed, by early 1928 he had managed what would turn out to be a personnel coup, hiring brilliant chemist Wallace Hume Carothers away from his teaching job at Harvard to head one of the research groups. In April 1930, a little more than two years after Carothers's arrival, chemists in his group produced both neoprene synthetic rubber and the first laboratory-synthesized fiber, which would lead to the development of nylon. These discoveries were among the most lucrative in the history of Du Pont.

Other changes were occurring, too. More large companies had begun to allocate at least a portion of their research budgets to advancing basic science. In addition, it had become increasingly common for businesses to draw on the resources of MIT; indeed, the Institute had begun to depend on the real-world grounding and financial benefits resulting from industrial ties. Despite this apparent merging of industrial and academic interests, cracks had begun to appear in some of these early attempts to unite industry and scientific research.

Willis Whitney of General Electric, for example, complained that the goal of creating an industrial lab to pursue basic research "has been impossible to maintain; the rule in GE has been to give calls for assistance from the engineers and production men precedence over all else . . . if they involve

. . . possible loss to the company or unsatisfactorily meeting a customer's needs."[13]

Similarly, although Wallace Carothers was able to make important contributions to polymer science during his years at Du Pont, the original concept of a pure science program was gradually dismantled. During the Depression, economic constraints and a new head of research began to profoundly alter the focus of Du Pont's fundamental research groups. In fact, it was under pressure from the new head to produce practical results that nylon was finally developed in 1934. Carothers's distress at this shift was clear. "The only guide we have for formulating and criticizing our own research problems is the rather desperate feeling that they should show a profit at the end," he wrote at this time. "As a result I think that our problems are being undertaken in a spirit of uncertainty and skepticism."[14] Carothers, who had suffered from bouts of serious depression well before he joined Du Pont, committed suicide in 1937.

## ACADEMIA AND INDUSTRY COLLABORATE

Just as industrial leaders who were graduates of the new engineering schools, such as the Du Ponts, sought to change industry, they also sought to build stronger ties to universities. The marriage was not always a happy one. Industry often felt that academia was not responsive to its needs, while academia complained that industry was too short-term oriented. In spite of the Morrill Act, many universities—including the scientific contingent that was beginning to be accepted on campuses—continued to look down their noses at institutions and disciplines such as engineering that stressed "practical" orientations. In 1891, Francis Amasa Walker, then president of MIT, wrote, "Too long have our schools of applied science and technology been regarded as affording an inferior substitute for classical colleges. Too long have the graduates of such schools been spoken of as though they had acquired the arts of livelihood at some sacrifice of mental development, intellectual culture, and grace of life."[15]

There were some very real incentives for industry to strike up a relationship with academia. The basic research undertaken by GE and Du Pont may have been impressive, but it was not indicative of efforts across industry. Although the benefits of industrial research—whether applied or basic—had become almost universally recognized, most companies didn't have the financial resources to hire or support an adequate scientific staff. It is particularly telling to note that in 1925, GE and AT&T together employed some 40 percent of the physicists working in industry. Most small companies, by contrast, had to look outside for research assistance.

To a certain extent, the federal government attempted to fill this gap. Such agencies as the Geological Survey, the Bureau of Standards, the Smithsonian Institution, the Bureau of Mines, and the National Advisory Committee for Aeronautics began to conduct some research, in part to aid industry and to make its operations safer and more productive. But these efforts were barely a start. World War I had temporarily boosted federally sponsored research, but in 1937 the government slated only 2 percent of its budget for research, with an emphasis on agriculture, not industry.

Nonetheless, representatives of both industry and academia increasingly began calling for a collaboration. "The very existence of the great research programs of industry is predicated upon the existence of a vast army of free, disinterested and even impractical researchers at work in the laboratories of colleges and universities," declared William E. Wickenden, an electrical engineering instructor at MIT who would go on to work at Bell Telephone Labs and Western Electric, and who eventually became president of the Case Institute of Technology in Cleveland.[16]

As envisioned, the benefits to industry of such a collaboration were clear: Companies, particularly those that could not afford their own research labs, could call on some of the top scientists in the country both for new ideas and for specific practical applications. The value of such a resource was potentially huge. Although the benefits to academia might at first glance have appeared purely financial, proponents argued that universities also stood to gain important practical insights from their industrial contacts. In particular, many universities began to respond to the corporate pressure to produce more well-rounded graduates by seeking out industrial input and involvement in their courses and in their research efforts.

Here, again, MIT was a leader. During the first three decades of the century, the Institute pushed the boundaries of what constituted legitimate academic research. Two factors in particular helped put MIT at the fore. Building on Rogers's founding philosophy, the institute's four presidents during this period championed the value of industrial ties. Moreover, as MIT graduates began to assume positions of prominence in the business world, these alumni ties provided a natural link between the Institute and industry. In the 1920s, for example, the chief executives of General Motors, General Electric, Du Pont, and Goodyear had all been classmates at MIT a few decades earlier. Although the cooperative attitude toward industry affected MIT overall, the electrical and chemical engineering departments developed the most prominent and innovative programs.

Electrical engineering became a separate department at MIT in 1902. Five years later, the department launched a policy of embracing industrial cooperation. Dugald Jackson, who took over as head of the department in 1907 and who led the cooperative effort, came to the job with a background merging academic and industrial experience. Moreover, Jackson was backed by a visiting and advisory committee—appointed the same year he took over—made up of top executives at major corporations, including Elihu Thomson of GE, and Charles F. Scott, chief engineer at Westinghouse and head of Yale's electrical engineering department.

With the committee's solid endorsement, Jackson quickly began soliciting research projects from industry. Early accounts included the Boston Chamber of Commerce and the Boston Edison Electric Illuminating Company. Although Jackson's motivations in launching the industrial research program included aiding industry and raising funds for MIT, he insisted that this practically oriented research would be directly beneficial to students, and that it would not deter the department from making advances in pure science. In 1910, Jackson wrote in a department brochure that "we are ready to undertake some of the more distinctively commercial investigations under the patronage or support of the great manufacturing or other commercial companies; but we hope to carry on the more important researches untrammeled by the limitations imposed by contributions of funds from commercial concerns."[17]

Jackson's efforts to push interaction between companies and the university did not stop with research. His dream was for electrical engineers to become the leaders of industry. To prepare his students for this goal, Jackson revamped the educational program in electrical engineering to include business and other nontechnical courses. In 1917, he established a cooperative education program with GE, in which students alternated six months of study with six months of practical experience at GE plants in Lynn, Massachusetts, and Schenectady, New York.

MIT was not the first school to offer a cooperative education program, having been preceded by the University of Cincinnati and Northeastern University, for example, but it became a model for what such programs can achieve. The collaboration with GE soon expanded to include Western Electric, Bell Telephone Laboratories, Edison Electric Illuminating Company of Boston, and Stone and Webster, Incorporated. By the 1980s, the network of corporations involved in this cooperative plan had spread nationwide to include Hewlett-Packard, IBM, Xerox, and Motorola, as well as such local companies as Digital Equipment and Analog Devices. MIT's chemical engineering department played a similarly pivotal role in pioneering industry–university cooperation, but in this case, the push to embrace business would prove far more controversial, as the department's leadership was split between two chemists with radically different visions of the Institute's role.

In 1903, the same year that MIT added its Graduate School of Engineering Research, Professor Arthur A. Noyes founded the Research Laboratory of Physical Chemistry. Noyes, who was known to make disparaging comments about the value of "teaching" engineering, envisioned MIT as a science-based university with a graduate school oriented toward basic research. His lab, for which he himself put up part of the money, was designed to pursue these goals, and it soon garnered an international reputation and produced the first Ph.D.s from MIT.

In 1908, however, a new lab was established, the Research Laboratory of Applied Chemistry, headed by Professor William H. Walker. Walker, who had returned to the Institute full time in 1905 after serving for several years as Arthur D. Little's junior partner, had distinctly different views about education than Noyes. "Science by itself produces a very badly deformed man," he once proclaimed, "who becomes rounded out into a useful creative being only with great difficulty and large expenditure of time."[18]

To avoid creating such "badly deformed" graduates, Walker eschewed pure science and instead designed his laboratory to solicit research contracts from industry. This applied science approach, he reasoned, would give students and faculty practical, real-life experience, and would give the Institute important industrial contacts. "There is with scientific men," Walker wrote in 1911, "a general awakening to the fact that the highest destiny of science is not to accumulate the truths of nature in a form no one but the select few can utilize, but that the search for truth can be combined with a judicious attempt to make the truth serve the public good. Thus the distinction which has existed between the terms pure science and applied science is rapidly falling away."[19]

By 1916, Walker appeared to be gaining the upper hand in the philosophical battle with Noyes. That same year, at the suggestion of Arthur D. Little, then chairman of the visiting committee of the department of chemistry and chemical engineering, Walker further strengthened the practical orientation of the Institute by founding the School of Chemical Engineering Practice. This program, which mimicked the cooperative education activities in electrical engineering, gave students a chance to alternate industrial experience with classroom training, and encouraged an increase in industrial contracts. The program was a rousing success, and during the 1920s Walker's successors at the school were responsible for some of the most important developments in the nation's emerging petroleum processing industry.

Not surprisingly, the diametrically opposed philosophies of Noyes and Walker led to long-term animosity. In 1919, MIT cast its vote for Walker's concept of technical education when President Richard Maclaurin asked Noyes to relinquish his role in the chemistry department. Noyes resigned and went to California, where he helped revamp the Throop Manual Training School into the California Institute of Technology. In this particular battle, industrial cooperation had won out over "pure" science. Walker's Laboratory of Applied Chemistry closed at the end of 1934, a victim of the Depression, but in the years following World War I, it conducted more sponsored research than all MIT's other labs combined.

A great deal of the collaboration was with Standard Oil of New Jersey (now Exxon Corporation). Under the leadership of Professor Warren Lewis, and later, Professor Edwin Gilliland, the department participated in an extraordinary relationship that lasted from the early 1920s to the mid-1970s. Not only did Lewis and Gilliland work closely as consultants on a wide variety of problems relating to the refinement and processing of petroleum, they even coordinated the founding of the Exxon Research and Development Laboratory in Baton Rouge, Louisiana. At Lewis's suggestion, the company hired Robert T. Haslam, then head of MIT's School of Chemical Engineering Practice, to assemble a fifteen-person task force of MIT students and staff to get the new lab on its feet. Six members of the task force, including Haslam, eventually became directors of Exxon or presidents or vice presidents of its major affiliates.

The battle in MIT's chemistry department highlighted the conflict over the proper role of the university in society and cast doubt on the basic philosophy behind the Morrill Act. Should universities really be concerned about educating the "industrial classes," preparing them to meet the specific challenges of their future jobs? Or are they seats of a "higher" learning that concerns itself more with the underlying principles of nature, human and otherwise?

This debate has frequently flared anew, and has never really been resolved. In the first part of this century, however, the clash of views proved to be highly productive. It led to the creation of new fields of instruction, such as chemical engineering. It contributed to the creation of independent industrial research labs. And it led to attempts by industry to work together with academia. Although many of these plans were successful, some failed outright. It is remarkable, however, that by World War II, engineers—a substantial percentage from MIT and other Boston-area schools—formed much of the top management of U.S. industry.

## FORMAL TIES

In 1918, MIT launched a pioneering program to establish a formal, direct link to industry. The incentive was primarily financial: For starters, the state legislature had refused to renew its land-grant appropriation, putting the pinch on the Institute's budget. In addition, George Eastman, founder of Eastman Kodak, had made an anonymous donation of $10.5 million to finance the construction of new buildings on the campus on the condition that the donation be matched by other private support. To spur contributions, President Maclaurin created the Technology Plan, the first systematic effort at MIT to coordinate and make the most of the school's growing connections to industry. Overseeing the plan was the newly created Division of Industrial Cooperation (DIC), directed by Walker, who passed on the mantle of the applied chemistry lab to take the new job.

In exchange for pledging support, companies received help from faculty and staff in keeping up with the latest research developments and in providing advice on industrial research programs as well as specific technical problems. In its first year, the plan raised almost $500,000, and pledges ultimately reached $1.2 million from 180 companies. President Maclaurin, declaring his support for the program, insisted—somewhat defensively—that, "It will not obstruct or interfere with the present educational necessities of the school. Education will not be sidetracked for commercial research. Cooperation will supplement education."[20]

Not everyone agreed with Maclaurin's assessment. Many faculty members charged that Walker was too aggressive in selling the Technology Plan and not thoughtful enough in choosing which research problems MIT was capable of handling. Partly in response to such faculty criticism, Walker resigned his post early in 1921. Other critics—setting out an argument that has dogged the Institute's contacts with industry ever since—claimed that the plan encouraged a too-cozy relationship between faculty and industry, and that both educational and research activities at MIT might be compromised.

Even more damning, however, was the corporate reaction. Most companies apparently did not feel they were getting their money's worth from the program, particularly when they were asked to pay research fees on top of the initial membership charge. Support for the program fell steadily after the first year, and in 1930 the plan was dropped. Despite its failure, the Technology Plan left behind an important legacy: a patent protection agency hired to represent the Institute, and the Division of Industrial Cooperation (later the Office of Sponsored Programs), an office experienced at handling contracts with companies.

For the next two decades, under President Karl Taylor Compton, MIT reevaluated its relationship with industry. Compton had been tapped for the job by Gerard Swope, an 1895 graduate of MIT and in 1930 president of General Electric and a member of the MIT Corporation. Swope and other members of the Corporation, including Bell Telephone Laboratories President Frank Jewett, felt that in order to keep up with increasingly sophisticated industrial needs, the Institute should strengthen its commitment to the sciences, which formed the underpinnings of all disciplines of engineering. Compton, a physics professor at Princeton, seemed to be the man for the job.

In 1930, in a conversation with Compton on a train headed for Washington, D.C., Jewett complained that engineering schools were in danger of becoming more like trade schools, and would decline in usefulness to industry unless they kept up with advances in basic science. Compton ultimately agreed to accept the post and directed much of his effort to building up the presence of science on campus. This focus indirectly helped to enrich the mix of the practical and the theoretical, boosting the sophistication of the field of engineering, and encouraging its broader acceptance in the academic world at large. Technology was becoming "high" technology.

Although Compton, who was president from 1930 to 1949, was a strong advocate of close ties between faculty and industry, he was also wary of the danger of becoming dependent on, or subservient to, industrial patronage. The failure of the Technology Plan, and the Depression-induced closing of the Research Laboratory of Applied Chemistry drove home the risk of relying on industrial contracts for research funds.

Moreover, there were more fundamental concerns to address. It had become clear during the 1920s that philanthropic organizations were bypassing MIT when doling out research grants. Many within the Institute insisted that this was because of the school's increasing emphasis on applied, rather than basic, research. In addition, while the Research Laboratory of Applied Chemistry had been impressively prosperous during the 1920s, its administrators complained that too often the projects undertaken were narrowly defined and of no scientific value. When there were significant findings, the industrial sponsors—as allowed in their contracts—often refused to let the lab publish the results. The very reputation for excellence that had originally attracted companies to the Institute was now in danger of being lost, as MIT's standing as a scientific institution was called into question.

As a result of these concerns, Compton immediately set out to redefine MIT's industrial interactions. The changes he enacted included barring industrial research that could be handled by private consulting firms, charging industrial sponsors stiff penalties to keep research confidential, and creating a new faculty promotion and compensation plan that rewarded original research. "Over-indulgence in consulting activities, to the detriment of Institute work is intolerable," Compton wrote in a 1932 Memorandum on Staff Personnel. "Whatever may have been the situation a generation ago we no longer have need for men who are primarily consultants and who incidentally conduct a class for us."[21]

Compton's changes had a profound impact. With the

tighter control on industrial contacts, MIT began to gain back whatever status it may have lost as a leading school of both engineering and science. Moreover, by reining in the Institute's business interactions, Compton reaffirmed its commitment to basic science as the primary goal of university research.

Ironically, a variation of the Technology Plan, allowing companies to stay abreast of on-campus research, surfaced in 1948—this time at the suggestion of industry. The resulting Industrial Liaison Program started up in August of 1948 with seven founding members: Standard Oil of New Jersey, Humble Oil, Standard Oil, The Texas Company (Texaco), Socony-Vacuum, Stone and Webster, and A. O. Smith Corporation. By the following year, Phelps Dodge, Union Carbide, Cities Service, Alcoa, Goodyear, and IBM had joined, and within five years, there were sixty member companies.

Unlike the previous plan, the Industrial Liaison Program caught on, and the idea gradually spread to most other research universities. For an annual fee, these programs typically provide information on current research, organize symposia on emerging areas of science, engineering, and management, and set up individual meetings between representatives of industry and academia. At MIT, the program had the effect of reinforcing the basic culture of cooperation by encouraging contacts among individual researchers from the academic and industrial communities.

## THE NEW INFRASTRUCTURE

Notwithstanding its subsequent philosophical and financial struggles, MIT had an impact on the development of the local business community within a few years of its founding. The Institute, combined with the existing intellectual, financial, and industrial strengths of the greater Boston area, increasingly attracted ideas and entrepreneurs to the area. A good seventy-five years before the opening of Route 128, Massachusetts already had sown the seeds of its high-tech community.

The telephone, for example, was an early innovation that came to fruition in Boston. When the young Scot, Alexander Graham Bell, first came to the Hub in 1871, his primary mission was to demonstrate how teaching the deaf might be revolutionized through the use of "Visible Speech," a symbolic representation of how the throat, tongue, and lips produce various sounds. Bell's approach was well received in Boston, where educators were in the vanguard of a movement to teach the deaf, rather than shut them away. His success resulted in his eventual appointment as Professor of Vocal Physiology at Boston University in 1873.

Bell's interest in and aptitude for acoustics went beyond teaching deaf children, however. In 1874, after hearing Bell lecture on Visible Speech at MIT, Professor Charles Cross, who had established the first course in electrical engineering at the Institute, offered him his technical advice as well as free access to MIT's apparatus and laboratories. Within two years, Bell had invented and successfully demonstrated the telephone.

Bell's telephone innovations and his work with the deaf were to remain inextricably linked. When Bell needed money to complete his early experiments, the fathers of two deaf children he had taught to speak—Boston attorney Gardiner Greene Hubbard and Salem leather merchant Thomas Sanders—helped out, and later put up the capital to form the Bell Telephone Company in Boston in August 1877. That same year Bell married Hubbard's daughter, Mabel, whom he had earlier taught to speak.

A plaque still marks the Cambridge building to which the

first two-way "long distance" call was made in 1876 from Bell's Boston laboratory. Sixty-six years later, that same Cambridge building served as the office and personal laboratory of Polaroid's founder, Edwin Land.

MIT's Professor Cross also inadvertently established an important and long-standing link between the Institute and General Electric. In 1883, a group of Lynn shoe manufacturers bought the Thomson-Houston Company of Connecticut, an electrical manufacturing firm, and moved the company's base of operations to Lynn, a Boston suburb. Soon after, Cross invited Elihu Thomson, founder of Thomson-Houston and a prolific inventor, to lecture to MIT's electrical engineering students. The bond that grew from that initial meeting became even more important when Thomson-Houston and Thomas Edison's New Jersey operations agreed to merge.

General Electric, the product of that merger, conducted much of the early research and development in electricity, and Thomson's continuing ties with MIT kept the two organizations closely allied. In fact, GE is a perfect illustration of the kind of close—and highly unusual—collaboration that existed between MIT and industry at this time. In 1899, Thomson served as one of MIT's first two nonresident professors. GE was the first company to participate in the Institute's innovative cooperative course in electrical engineering, and when MIT President Richard Maclaurin died unexpectedly in early 1920, Thomson served as acting president for almost three years.

The give and take between MIT and industry during this period led to even closer collaborations. Although by the 1970s, professor-founded high-tech companies had become almost a cliché in Massachusetts, early in the century the concept was still largely unknown. One man in particular, electrical engineering professor Vannevar Bush, proved the potential of this kind of collaboration.

Bush's first venture into business, as the head of technical and research facilities for the short-lived American Research and Development Corporation (AMRAD), a manufacturer of radio sets, transmitters, and components backed by the powerful J. P. Morgan, Jr., was not particularly successful. But when most of the same principals, including Bush, put together a new venture a few years later, they struck gold. Raytheon, originally dubbed the American Appliance Company, set up shop in 1922 as a refrigerator manufacturer on the third floor of what was then the Suffolk Engraving Company in Kendall Square.

Bush and Laurence K. Marshall, a Raytheon co-founder, had been roommates at Tufts University at the turn of the century, and had often challenged themselves by taking courses deemed too tough for undergraduate engineers. Bush, the stern and disciplined academic, eventually rose to the position of vice president and Dean of Engineering at MIT. Marshall, much more the risk-taking visionary and energetic promoter, was a natural entrepreneur.

Something of the essence of both men was captured in Raytheon, a new kind of company that brought together the skills of researchers such as Bush with those of businessmen such as Marshall. Furthermore, the establishment of Raytheon both marked and exemplified the beginning of a pioneering partnership between academia and industry that led to profound changes in the region.

Marshall was ahead of his time in seeing the importance of new scientific knowledge to opportunities in the marketplace. In an era when the federal government still had little interest in basic research, and only a tiny handful of the largest corporations had laboratories, Marshall plowed large sums of the small firm's money into both basic and applied research. As a central part of his strategy to stay at the leading edge, he collaborated with faculty at MIT's departments of electrical engineering, physics, and mathematics. This not only allowed him to keep up with the latest progress in a variety of key fields, but helped keep his own labs supplied with top-notch graduates.

Researchers from Raytheon and MIT often worked side-by-side on specific projects. This joint research ranged from work on ultrahigh radio frequencies and dielectrics to aircraft guidance and detection. One group designed and built the special shortwave communication devices that Admiral Richard E. Byrd used in a series of Antarctic expeditions beginning in 1929. Gradually, the company developed an extensive network of consultants, collaborators, and contacts at MIT.

According to Charles Francis Adams, former president and chairman of Raytheon, these ties were critical. "Our competitive survival has depended on our exploiting those links," he declared in a recent interview.[22] In *The Creative Ordeal,* an account of Raytheon's history, author Otto Scott points out that the special atmosphere of cooperation among MIT's theoreticians and Raytheon's practical engineers "extended the firm's position to a unique scientific and engineering promontory beyond the usual commercial enterprise."[23] The company became a pivotal player during World War II, and eventually grew to be the state's second largest employer and a leader in the nation's defense industry.

Not all the Boston area's early "high-tech" start-ups prospered, of course. For example, no trace is left of the automobile manufacturing industry that thrived in the area around the turn of the century. Perhaps the best-known product of this era was the Stanley Steamer, an early steam-powered automobile designed by twin brothers Francis E. and Freelan O. Stanley from Maine. Boston's auto industry lost out to another part of the country, just as the textile and shoe industries closed up or moved away several decades later. The company that made the Steamer—along with the more than one hundred other car manufacturers that sprang up in the area—eventually was eliminated by Detroit's faster-starting cars, mass production techniques, and marketing savvy.

Other companies were less "high tech" but demonstrated the crucial role that technology was to play in broad product areas. In the spring of 1895, King Gillette, a traveling salesman for a bottlecap company and something of a political crackpot, came up with an idea that was to change his life as he was shaving one morning in Boston's posh Back Bay district. Why not make a thin razor blade, he reasoned, that could be mounted in a holder that would guard against the deep cuts that were all too easy to inflict with a straight razor? When the thin, cheap blades became dull, the user could toss them out and replace them instead of taking the time and effort to sharpen them.

The key to Gillette's scheme was the ability to reliably and cost effectively mass-produce the small, sharp blades. Almost six years later, Gillette convinced William Emery Nickerson, an MIT-trained chemist with a reputation as a talented inventor but a bad businessman, to take on the challenge. Over a period of several months in a small shop near the waterfront, Nickerson drew on his knowledge of materials and manufacturing to devise the machinery that could harden and sharpen the blades rapidly and cheaply. The invention was a success, and the Gillette Company made the fortunes of both men.

By the 1930s, Boston was beginning to produce more than its share of small, innovative companies. Pioneers such as

Arthur D. Little and Raytheon had been joined by a spreading web of new technology companies, such as Polaroid and Foxboro. The founders of some of these companies were drawn to the area specifically to be a part of the burgeoning academic–industrial complex.

Moreover, in the labs of MIT ideas were being developed that in the wake of World War II were transformed into new products and companies. Robert Van de Graaff built his first high-voltage accelerator, a machine that lay at the heart of High Voltage Engineering's founding in 1946, and which is still on display in Boston's Science Museum. And professor and inventor Harold "Doc" Edgerton joined with research assistants Herbert Grier and Kenneth Germeshausen in an informal partnership that would later become EG&G. It took the intense effort, focus, and infusion of federal funds of the war years to ignite Boston's high-tech community, but a network had already begun to evolve.

Another key element of the infrastructure was under rapid development at the time as well: the academic–medical complex. For example, the Harvard Medical School, founded in 1782, was steadily extending its network of affiliated patient care and research facilities. Again, the roots of this complex lay in the practical tradition of the region merging with the intellectual.

The vast majority of university research then, as now, did not result in a specific product, process, or company. Campus research, after all, was intended to advance basic understanding of physical phenomena. There were some striking exceptions, however, cases where an effort to push science forward led to unexpected, long-term results.

One particular example stands out. Soon after his appointment, President Karl Taylor Compton, along with others at the

Institute, began pressing for a more probing examination of unexplored aspects of biology. In a paper he presented to the American College of Physicians in 1938, Compton declared that the challenges of biology "are so complex, so vast in scope and human import, that every promising avenue of approach to their solution should be followed."[24]

Compton's intentions fit neatly with those of Warren Weaver, then director of the Rockefeller Foundation's biology program. Weaver, who advocated an "exact, analytic, vigorously formulated"[25] biological research agenda based on the methods of physics and chemistry, sought out and funded a select group of top scientists at a few institutions, including MIT and Harvard, allowing them to bring a new perspective to an age-old discipline. This fundamental enhancement of the direction and nature of biological research led to the development of the field of molecular biology and, ultimately in the 1970s, to the explosive field of genetic engineering.

Despite his concerns about how MIT should relate to industry, Compton was by no means antibusiness. Under his administration, MIT continued to forge strong industrial links. The Division of Industrial Cooperation (DIC) had become well established, and—in contrast to most universities—the process of setting up and administering contracts with industry had become old hat. MIT had made it clear that it wanted to interact with outside institutions, and that it knew how to do it.

Compton encouraged another new idea that ultimately revolutionized the high-tech world. In the 1930s, he was chairman of the New England Council, a regional business association founded in 1925 whose members included government officials, business leaders, and a few academic leaders. At the time it was already clear that New England's industrial base—

including textiles, leather, and machine tools—was on the decline as other regions with cheaper labor costs drew away manufacturing jobs. Simultaneously, new businesses based on advances in technology were beginning to dot the landscape: Bell Telephone was already a giant, General Electric was thriving, and dozens of other companies were struggling to get off the ground.

Recognizing the economic potential of these new firms, Compton and his colleagues hit on the idea of forming an institutional venture capital firm. Its purpose would be to provide investment funds to nurture the development of new high-tech companies, and thereby encourage the diversification of the regional economy. Although the idea seemed a good one, the effort was stalled by the outbreak of World War II.

With the coming of the war, MIT was in a special situation. A number of factors contributed to its role as the primary American center for war research, and the Institute's openness to outside interactions and its ability to manage contracts helped make it a natural choice for federal military research and development. In fiscal 1940, the Division of Industrial Cooperation administered less than $100,000 worth of contracts. By September 30 of the next year, MIT's research contracts, mostly with the federal government, had topped $10 million. The war years had begun.

## MAKING CONNECTIONS

The infrastructure that arose in Massachusetts before World War II was profoundly influenced by MIT and its willingness to interact directly with the outside world. The wide range of institutional and faculty involvements that ensued proved remarkably productive. New, practical ideas were being conceived. New links to industry and new companies were being formed to exploit them. New technologies were moving to the marketplace. And the idea of formalized venture capital was born to facilitate the process.

Yet few of these developments were easy. In fact, it could also be argued that the Morrill Act began a face-off among industry, government, and academia that has lasted to this day. Government wasn't happy providing long-term subsidies to the states, and eventually slacked off as wealthy industrialists and their foundations fortuitously picked up the torch. Industry was never quite satisfied with the focus of university research, nor with the skills and capabilities of graduates.

For its part, academia was thrown into turmoil. Scientists were accepted into the academy with relatively little dispute, but the far more practically oriented engineers were regarded with suspicion and even hostility, not only by the traditional academicians, but also by the recently accepted scientists. The engineers' ongoing contacts with industry—an interaction that tested the established boundaries between academia and industry—exacerbated the feelings of difference.

Despite these tensions both within and among the various sectors, it is clear that the first time government, industry, and academia came up with a plan to work together, they hit a home run. Arguably because of the Morrill Act's evolution, on the eve of World War II, the United States had an extraordinary network of research universities that had been turning out technically trained talent in unprecedented numbers. Massachusetts, with its high concentration of academic institutions, its industrial heritage, its emerging high-tech infrastructure, and its connections to the sources of power in government, was set to play a key role in future developments.

# 3

# The Government Steps In: The Legacy of World War II

ON JUNE 15, 1940, the day after the German war machine thundered into Paris, Vannevar Bush walked into President Franklin D. Roosevelt's office with a novel plan for marshaling the nation's top scientific and engineering talent in the event that the United States entered the war. (Bush had moved to Washington, D.C., in 1938 to serve as president of the Carnegie Institute and chairman of the National Advisory Committee on Aeronautics.) He outlined his proposal in four short paragraphs typed neatly on a single sheet of paper. Just ten minutes later, he walked out with "OK-FDR" scrawled across the top.[1]

Eighteen months later, in December 1941, the United States declared war on Germany and Japan in the aftermath of the attack on Pearl Harbor. By then, Bush's plan was already well underway. Neither he nor Roosevelt suspected the overwhelming impact it would have, not only on the course of the war but also on the subsequent technological dominance and economic strength of the nation for decades to come. Bush's plan also helped lay the groundwork for the postwar surge to prominence of Boston's high-tech community.

World War II not only spurred the development of a wide

range of technologies, it also introduced the federal government as the permanent and principal sponsor of basic research—for both military and civilian interests—in the United States. It spawned a vastly expanded research community integrating universities, industry, and government laboratories. And it set the scene for an explosion of start-up companies, the rapid growth of existing ones, and the birth of entirely new industries.

For the Boston area, World War II had especially pervasive consequences. Because of the unusual academic and research community that had already developed there, and its contacts with both the military and industry, the region was exceptionally well prepared to capitalize on developments during the war and in the years following. The seeds Bush planted had far greater impact than he could have imagined. It is fair to say that the federal presence in Massachusetts, which began in earnest during World War II, catalyzed the technological strengths of the region, leading to the commercialization of emerging technologies not just by local companies, but also by a number of major corporations outside the region.

The development of advanced microwave technology, the construction of the first practical digital computer, and the creation of guidance systems for space are but a few of the Boston area's technological achievements backed by defense dollars. Introduced to the strategic importance of science and technology, the military became an ardent and continuing supporter of research and development. More to the point, it became an extraordinarily wealthy customer that demanded just the sorts of expertise and services the Boston area was equipped to provide. The result was an infusion of funds that helped both academia and industry flourish.

Military contracts caused the local R&D community—including MIT and other universities, companies such as Raytheon and EG&G, and new government laboratories—to soar in size. This, in turn, provided the means to train and employ even more scientists and engineers in critical areas of technology. Moreover, government contracts allowed aspiring entrepreneurs to take their expertise to the marketplace. By supporting leading-edge research, the government not only encouraged progress in a variety of fields, it also helped push innovative ideas closer to the point of commercial viability.

## ACADEMIA MAKES AN END RUN

Bush's pivotal meeting with FDR was the culmination of a number of informal discussions about the alarming developments in Europe among several of the nation's academic and scientific elite, most of them members of the National Academy of Sciences. Many of these leading scientists and engineers believed that technological superiority could easily tip the scales in favor of one side or the other in the event of war. The 1914–1918 war, they recalled, had caught the nation off guard. German U-boats had wreaked havoc on shipping, and the nation's war research effort had been slow to start up.

Most research then had taken place at the network of government labs based primarily at arsenals around the country. Progress was slow, and separate research groups often worked on the same problem, unaware of each other's existence. Moreover, the government's poor communication with industry hampered the process of turning new ideas into workable and reliable mass-produced devices. Bush had seen some of these problems firsthand. Not long after the United States entered the war, he had developed a simple and desperately needed technique for detecting the deadly U-boats that lurked in the safety of the ocean's depths, sinking Allied surface ships almost at will. However, Bush had such a tough time penetrating the Navy's bureaucracy that the war ended before the device could make it to the scene of action.

In the years after 1918, new technologies blossomed. There were important advances in such basic fields as physics and chemistry, as well as in more applied fields such as electronics and medicine. Bush and his colleagues did not believe, however, that the military research establishment of the 1930s had the expertise needed to exploit the basic research that held such promise. In intense discussions, the group shared its concerns about the precarious state of American weapons technology, and its fears that the nation's legislators and military establishment could not move quickly enough to take advantage of available ideas.

Bush came up with his plan in the spring of 1940. Why not, he argued, set up a small, independent group reporting only to the President to guide and coordinate new areas of research? This approach would allow innovative research to proceed unencumbered by the military bureaucracy and would complement work at existing military labs. And why not take advantage of the expertise and facilities already available at the nation's universities? Such an "independent" organization, he reasoned, would be able to coordinate both industrial and military resources efficiently.

Perhaps in part because the plan placed control in their own hands, Bush's proposal easily won the approval of his colleagues, who included James Conant, a chemist and president of Harvard University; Karl Compton, physicist and president of MIT; Frank Jewett, then president of the National Academy of Sciences, and the man credited with building Bell Labs into the world's most renowned industrial research lab; and Richard Tolman, a theoretical physicist and dean of the California Institute of Technology. Although Bush, a straitlaced, conservative Republican, did not seem the natural choice to deal with a Democratic administration, the group chose him to carry the idea to the President because of his Washington connections. Just a few months after his meeting with Roosevelt, the proposed National Defense Research Committee (NDRC) was a reality.

In subsequent months, conflict in Europe escalated. Bush moved quickly to bring the atomic advisory committee under his control and, heading off the efforts of the American Medical Association, which he and his friends distrusted, he took over the supervision of plans to conduct wartime medical research as well. A new entity, the Office of Scientific Research and Development (OSRD), with Bush as its head, was created in the spring of 1941 to oversee both the NDRC and the newly formed Committee on Medical Research.

As he and his cohorts had planned, Bush reported directly to the President. OSRD did no research itself, but it determined what specific research should be done, and who should do it. Bush administered the funds appropriated by Congress on the recommendation of committee members, awarding contracts to university, private, or governmental laboratories. And although OSRD sought the approval and cooperation of the Army and Navy on its projects, it had the authority and the means to pursue them no matter what the armed forces said.

Naturally, not everyone was pleased with the extraordinary setup, but Bush made no apologies. Recalling the period in his memoirs, *Pieces of the Action,* he wrote:

> There were those who protested that the action of setting up N.D.R.C. was an end run, a grab by which a small company of

> scientists and engineers, acting outside established channels, got hold of the program of developing new weapons. That, in fact, is exactly what it was. Moreover, it was the only way in which a broad program could be launched rapidly and on an adequate scale. To operate through established channels would have involved delays—and the hazard that independence would be lost, that independence which was the central feature of the organization's success.[2]

Although controversial, Bush's plan for bringing together the talents and facilities of the military, industry, and universities for a common purpose was inspired. In practice it appeared to be a natural alliance. Working under the auspices of OSRD, and with the unifying motivation of the war, thousands of scientists, engineers, managers, production workers, and military personnel designed, built, tested, mass produced, and delivered a vast array of new devices and techniques for waging war. By the end of the war, OSRD had spent more than $425 million in research and development.

OSRD was not the only recipient of federal defense funds. Millions of dollars were also spent at existing military labs, but that research was aimed primarily at perfecting and improving conventional military systems. OSRD projects were more dramatic. Among the Office's achievements were a variety of radar systems; more deadly artillery and antiaircraft shells that exploded as they sensed they were near a target; amphibious vehicles that proved indispensable in the Normandy invasion; improved drugs for the treatment of malaria; a broad array of new antibiotics; self-aiming antiaircraft guns; and the atomic bomb.

Never before had government, industry, and academia merged their talents on such a scale, and never again would the relationship between the participants be the same. Without fully realizing the long-term impact of his proposal, Bush had served as midwife to the birth of the modern academic–military–industrial complex. Moreover, the government was now firmly ensconced as the country's primary sponsor of basic research.

Probably no other state benefited as much from Bush's redirection of government research spending—and the commercial spillover that resulted—as Massachusetts. And without a doubt, no university reaped more rewards than MIT, which became the nation's unofficial center for wartime research. Although a number of other key universities took on OSRD contracts, including the California Institute of Technology, Columbia University, and the University of Chicago, MIT was the largest contractor by far, garnering a total of seventy-five contracts representing nearly $117 million. Harvard, in part because of the government's desire to locate additional programs near MIT, ranked third after Cal Tech, receiving seventy-nine contracts worth $31 million.

MIT topped the list for several reasons. Obviously, it didn't hurt having Bush at the head of OSRD. With his many MIT connections, it was perhaps inevitable that Bush would think first of his former employer when delegating contracts. This same "old boy network," which extended throughout the academic and industrial research communities, helped MIT graduates dominate the Office's various committees. There were also practical and compelling reasons for siting major projects at MIT. In 1940, the Institute was recognized as one of the nation's leading schools in many areas of science and engineering. Its faculty and students were already doing pioneering work in technologies—such as microwaves—that were to be crucial in the war, and its research and educational programs were strongly oriented toward practical results. In addition, it was more efficient to cluster several projects in one location, drawing on a number of disciplines simultaneously, than to scatter them all across the country.

MIT was also well suited for the collaboration with industry that the wartime effort demanded. Unlike most other universities, MIT encouraged its faculty to consult with and to consider the needs of the business community. The Institute also had strong ties to strategically important industries as a result of its disproportionately large share of alumni employed in key technical positions. It has been estimated, for example, that 70 percent of Princeton's alumni were in uniform during the war, but only about 30 percent of MIT's graduates served in the military, contributing instead as researchers in industry, government, and academia.

The importance of MIT's wartime role to the development of the region's high-tech community can hardly be overstated. MIT had already laid down a foundation of innovation and technical talent on which to build, and the mobilization effort brought in skilled collaborators from government, industry, and other universities; set forth new, compelling challenges that needed solutions in record time; and provided the cash necessary to leapfrog previous achievements and to usher in a new generation of technologies.

When the war ended, many of the scientists and engineers drawn to the region stayed on. The technological developments that had proved so important to designing new weapons and defense systems proved equally useful to the commercial marketplace. And the federal research dollars continued to flow. As the following stories illustrate, the war years gave rise to new technologies, new companies, and new industries that helped to position Boston at the forefront of technological innovation.

## THE MARKET FOR MICROWAVES

The largest single program at MIT, and the largest wartime effort outside of the Manhattan Project, which led to the atomic bomb, was the effort to perfect radar. While the British had developed the fundamental concepts of radar and begun to employ them in defense of their nation, existing devices were still fairly unsophisticated when the war broke out in Europe; they could do little more than detect the presence of activity in the air.

Nonetheless, the rudimentary system gave the Royal Air Force a fighting chance against the superior numbers of the Luftwaffe, Germany's air arm of bombers and fighter planes. An alarming problem surfaced, however, when the British realized that they needed more radar setups and did not have enough machinists in the entire country to make a sufficient number of magnetrons, the powerful microwave generating tubes at the heart of the systems. Magnetrons had to be whittled out of solid chunks of copper by skilled machinists, a tedious and labor-intensive process.

Desperate for help, the British decided to let the United States in on the top secret design of the magnetron. A team of scientists brought one to the United States in a nondescript black box in September 1940, hoping that they could enlist the aid of Bell Labs and its manufacturing arm, Western Electric, in the design and production of more and better devices. NDRC had already set up a Microwave Committee, headed by MIT president Karl Compton, to oversee research in the field. Members of the committee included representatives of Bell Labs, Sperry, RCA, General Electric, and Westinghouse Electric Corporation, assembled to pool their engineering and production expertise. Some of these companies already had ties to MIT, which had established itself as a leading center in the field. The black box was opened at a meeting in Washington, D.C., revealing a device that, though difficult to make, was nearly 1,000 times more powerful than any microwave generator developed in the United States.

Impressed with the British device and its potential in the microwave field, the Microwave Committee moved quickly, setting up a facility at MIT in October 1940 to coordinate further radar and microwave research. They named it the Radiation Laboratory to mask its true purpose. Casual observers would be led to think that the lab focused on nuclear physics, which, ironically, had no apparent military application at the time. When the United States at last joined the war after the attack on Pearl Harbor just over a year later, radar research was already in full swing.

The researchers from academia and industry who came together at the Rad Lab had already ironed out much of the basic theory on the nature of shorter wavelength radio waves. Project leaders were faced, however, with the prodigious task of engineering practical devices based on the scientific concepts, and then mass producing the finished product. This required the coordination, on a grand scale, of scientists, engineers, managers, manufacturers, and the ultimate users—the military. In an unusual twist of conventional management practice, scientists and engineers had complete control over the course of systems development, and their efforts were served and supported by representatives of the military and industry.

At its peak the Rad Lab had a staff of over 4,000 and filled over fifteen acres of floor space around Cambridge. It was an extraordinary concentration of talent, whose meshing proved a fertile union. The Lab developed over 150 systems that applied the versatile microwave technology to a dizzying array of applications, from air and sea navigation to the detection of enemy aircraft and submarines and even to improving the accuracy of bombs and artillery fire. Toward the end of the war, it was not uncommon for a B-29 bomber, for example, to have as many as five separate radar systems aboard.

As the Rad Lab was being launched, a small but aggressive local company responded to the Microwave Committee's appeal for a solution to the problem of speeding up the magnetron production process. Raytheon, then a radio tube manufacturer with only $3 million in annual sales, was a midget compared to the powerful members of the Microwave Committee. But co-founder Laurence Marshall, intrigued by the technology's commercial potential, had already developed an interest in microwaves before the war.

Although the British were at first reluctant to share their nation's most closely guarded technological secret with a tiny firm they had never heard of, the connection proved to be fruitful. Raytheon engineers soon came up with a novel way to boost production of the devices dramatically by assembling them out of laminated copper sheets instead of carving them laboriously out of solid blocks. Experts had predicted that it would be impossible to turn out more than 100 magnetrons a day because of the complex machining involved, but Raytheon's technique eventually enabled the firm to turn out over 2,000 a day.

This ability to make magnetrons relatively cheaply and quickly spurred the further development and application of the technology. Toward the end of the war, Raytheon commanded over 80 percent of the magnetron market. In addition, the company took advantage of its proximity to MIT to work with

Radiation Laboratory scientists and engineers on a number of other radar-related components and systems.

The wartime effort transformed the little radio tube company. From 1940 to 1945, Raytheon's sales skyrocketed from $3 million to more than $150 million as the number of employees jumped from just 1,400 to some 16,000. In 1991, Raytheon was the state's largest defense contractor and second largest company, and was reveling in its role as developer of the Patriot missiles, the antiballistic missiles the United States used so effectively against Iraqi SCUDS during the Gulf War.

Raytheon was only one of many companies whose fortunes were made by association with the Rad Lab. A new industry sprang up in the space of just five years. By 1945, American companies had delivered nearly $3.1 billion in radar equipment to the armed forces. And in the years following the war, continued demand for more and better radar systems, coupled with new uses for microwaves in communications, led to over $100 million in magnetron sales alone between 1945 and 1950.

## PUTTING COMPUTERS TO WORK

Without a doubt, World War II radar systems research gave a powerful boost to the electronics industry, especially in Massachusetts, but a much smaller wartime effort was to have a far greater impact on the region, and the nation as a whole. Thanks largely to an obscure project begun late in the war, Massachusetts unexpectedly came to play a crucial role in the development and commercialization of computers. Central to this development was a landmark achievement in technology funded by the military: the world's first fast and reliable digital electronic computer.

The concept of a machine that could perform tediously repetitive calculations rapidly and accurately had long been a challenge to the academic community. By World War II, a few bulky and elaborate electromechanical devices existed that could be made to grind through simple operations, but these machines were large, expensive, and extremely inflexible. For example, the Mark I electromechanical computer developed by Howard Aiken, the irascible director of Harvard's Computation Lab, was over fifty feet long and eight feet high, and weighed more than five tons.

Serious work was underway at several other institutions, particularly the University of Pennsylvania, where John Mauchly and Presper Eckert were working on the Electronic Numerical Integrator and Computer (ENIAC). This machine, funded by the U.S. Army Ballistics Research Center, was unveiled in February 1946. ENIAC's vacuum tubes made it a speedier machine, but it was still inflexible and it had a severely limited memory.

It was MIT that managed to leapfrog these earlier, more primitive, devices with a machine that ushered in the modern age of computers. In 1944, Captain Luis de Florez, a graduate of MIT and director of the Special Devices Division of the Navy's Bureau of Aeronautics, sought the advice of his alma mater on the feasibility of building a general-purpose aircraft "simulator" that would reduce the cost and time of training pilots. The British had already put the idea to work with simple pneumatic systems, each built to imitate specific aircraft and Bell Laboratories had developed several similar electromechanical systems. De Florez's idea was to take the concept a step further by devising a system that could be "programmed" to imitate more than just one aircraft. The response to the controls would be dictated by an "analyzer" that would calculate the proper dynamic response to the pilot's commands based on its stored knowledge of the characteristics of the airplane being simulated.

To MIT researchers, such a general purpose system—dubbed the Airplane Stability Control Analyzer (ASCA)—held intriguing possibilities as a research tool as well as a trainer. It could be programmed to simulate proposed craft as well as existing ones. Using data from calculations and wind tunnel evaluations of new and untested aircraft designs, they reasoned, it should be possible to allow experienced pilots to "fly" new aircraft before they were built. The pilots' added input could then greatly speed up the process of developing new airplanes.

Since Bell Labs had already built some similar, simpler devices, de Florez and experts at MIT felt that Bell would be the logical chief contractor for the proposed universal simulator. With the war winding down, however, both Bell Labs and Western Electric shied away from committing to a project of seemingly limited commercial significance. Besides, they were both still loaded down with a number of other wartime projects.

Instead, MIT's Servomechanisms Laboratory accepted the task. The Lab, working with such firms as the Sperry Gyroscope Company, had spent the war developing a variety of dramatic "servomechanisms," gadgets that, among other things, could automatically control antiaircraft fire from ships by performing calculations that took into account the speed and trajectory of the target as well as the specific characteristics of the gun.

Jay Forrester, a lanky electrical engineer from Nebraska, was assigned to the problem and attacked it with characteristic zeal. It was not an easy task. In particular, the simulator's "brain" had to be able to receive the pilot's commands, zip through the complex calculations that would tell how the plane should respond, and then tell the machine how to act—all in just a fraction of a second. Although no existing machine could come close to those requirements, Forrester firmly believed it could be done. Moreover, he realized that if it could be done, it would have applications far beyond training pilots.

It was soon clear that the program would take more time and effort than originally planned. In his progress reports to the Navy, Forrester kept insisting that the "brain" itself had far broader potential both as a research tool and as a part of a wide variety of military command and control systems. All pretense of building a simulator was abandoned when the researchers finally tossed out the cockpit in June 1948. The evolving computer was named "Whirlwind," and it became one of several computer projects being funded by the Navy's Special Devices Division.

The Navy was gradually becoming disenchanted, however, as Forrester's budget requests skyrocketed, deadlines lapsed, and new technical challenges popped up. It also had some difficulty grasping Forrester's vision. To the Navy supervisors, computers were tools for mathematicians, and while they saw the value of computing machines for supporting work in advanced technology, they weren't sure how they could be applied in practical military systems, as Forrester was maintaining. Furthermore, some of the Navy's other computer projects, such as the one being conducted by eminent mathematician John von Neumann at Princeton's Institute for Advanced Study, were not nearly so expensive.

Reluctantly, the Navy continued its support, and Forrester forged onward with his band of students and staff researchers. If it had not been for the intervention of events far beyond the borders of New England, however, the Whirlwind project might have been scrapped. In the years following World War II, America's relationship with the Soviet Union deteriorated rapidly as Cold War tensions mounted. The schism widened when the Russians entered the nuclear age in August 1949 by detonating an atomic bomb. And in the fall of 1950, the Korean War broke out. Members of the military and scientific communities gradually realized that continuing advances in aircraft and missiles could for the first time make the United States vulnerable to air attack.

Defending thousands of miles of border from such an attack was a technical problem of staggering proportions, and the existing detection system, which relied on individual radar operators to interpret data from each radar station, was too slow and inaccurate to be of use in a surprise attack. Once again, the old boy network tying MIT to the government and the military went into effect. A committee headed by MIT physics professor George Valley, a member of the Air Force Scientific Advisory Board, produced a report in October 1950 calling for an automated air defense system consisting of a coordinated network of radar stations strung along hundreds of miles. But no system existed that could analyze the data fast enough.

Two months later, Air Force Chief of Staff General Hoyt Vandenberg, recalling MIT's contributions through the Rad Lab in World War II, wrote a letter to MIT president James Killian, asking the Institute to set up a laboratory to conduct the research and development required to design and build a new national air defense system. While Killian was weighing the matter, he went to Washington to meet with Dr. Louis N. Ridenour, formerly assistant director of the Rad Lab and then Chairman of the Scientific Advisory Board. Trying to convince

Killian of the plan's merits during a walk in Lafayette Park, Ridenour prophetically insisted that the proposed lab would make Massachusetts a center for electronics research and industry. Killian ultimately acquiesced, and the Institute set up Lincoln Laboratory, a nonprofit organization operated by MIT for the Air Force.

Professor Valley, who had made the original appeal for an automated air defense system, monitored Project Lincoln's progress with interest. While walking across campus one day in early 1952, he ran into Jerome Wiesner, an electrical engineering professor who later became president of MIT. Valley confided to Wiesner the difficulty of correlating vast amounts of radar information quickly and reliably. Wiesner suggested he check out Jay Forrester's work. Valley was so impressed with the potential of Whirlwind for the task that, less than a month after visiting Forrester's lab, he assured him that the Air Force would pick up the tab for his work.

Snatched from the brink of disaster, the Whirlwind project finally had a clear sense of purpose agreeable to both Forrester and his new sponsor. Moreover, the work already completed set the stage for rapid progress. By 1953, Whirlwind was operating reliably as the heart of the "Cape Cod System," a prototype air defense network consisting of a small string of radar stations in the Cape Cod area, successfully tracking as many as 48 aircraft simultaneously and providing guidance to intercepting aircraft.

The promise of the approach helped convince the Air Force to shut down a competing project under development at the Willow Run Research Center of the University of Michigan and focus its resources on the Lincoln system. The cancellation was a coup for Massachusetts. Politicians in both regions had lobbied the Air Force, believing that such work would bolster their regional economies. The yes vote for Lincoln also bolstered the fortunes of IBM, which had begun collaborating with MIT on the project the previous year.

In January 1955, IBM delivered to Lincoln Lab the XD-1 computer, patterned after the Whirlwind, to serve as the "brain" of what came to be called the SAGE (Semi-Automatic Ground Environment) air defense system. Other firms that developed parts of the system included Bell Labs, the Western Electric Company, the General Ceramics and Steatite Corporation, General Electric, Convair, Hughes Aircraft, and Raytheon.

By the end of that year, Forrester's project could finally be judged a clear success. Whirlwind I was up and running an average of more than 97 percent of its 168-hour work week. Critical to its performance was a new, fast, and compact electronic memory first sketched out by Jay Forrester in 1949. The concept involved tiny doughnuts of magnetizable material strung into a three-dimensional array of wires. This so-called "magnetic core memory" concept was not only key to Whirlwind's operation, but it later formed the heart of IBM's System 360, the computer that first catapulted the firm into the position of dominance in the industry it still holds today. After a bitter licensing dispute with IBM, MIT's patent on the concept went on to become the most lucrative it has ever held, bringing in a whopping $22 million over its seventeen-year lifespan.

The Whirlwind project's greatest legacy, however, may have been spawning the company that created the minicomputer industry in Massachusetts, and that, ironically, grew to be IBM's chief rival in the multibillion-dollar computer industry. Accepted to work in Forrester's group, Ken Olsen was one of the engineers who helped perfect the memory concept. His master's thesis in 1952 involved a novel electronic switch design for selecting a specific memory site. After graduating, Olsen moved to Lincoln Lab to work on the further development of the SAGE computer system. In 1957, convinced that computers had tremendous commercial potential, Olsen left to found Digital Equipment Corporation.

Another future heavyweight in the industry, Wang Laboratories, had already spun off from Harvard's Computation Lab in 1951. Founder An Wang, a native of China who had earned his Ph.D. in applied physics from Harvard in 1948, planned to work at Hughes Aircraft, but instead took a job at Howard Aiken's Computation Lab rather than fill out all the security clearance forms required by Hughes. Wang knew little about computers, but he felt the work might be interesting. For his first assignment, Aiken asked Wang to figure out how to make a computer memory that had no moving parts. At the time, memories tended to rely primarily on inherently slow mechanical or electromechanical components. Within three weeks, Wang came up with the idea of a magnetic core memory. In fact, the technical papers written by Wang inspired Forrester to arrange Whirlwind's magnetic cores in a three-dimensional array.

## RESEARCHERS FOR HIRE

Olsen and Wang were not alone in seeking to apply their technical expertise in the business world. In fact, a high-tech explosion was in the making in the 1950s and 1960s, fueled partly by military money. In a series of theses coordinated by Professor Edward Roberts of MIT's Sloan School of Management, researchers identified 129 Boston-area companies launched in the postwar period by entrepreneurs associated with just three of MIT's academic departments.

The military's keen interest in maintaining a technological edge, whetted by the experience of OSRD during the war, was sharpened further by the Soviet Union's detonation of its first atomic bomb in 1949, the Korean conflict in the early 1950s, and the successful launch of Sputnik by the Soviets in 1957. As a result, funding for R&D in leading-edge technologies soared, and the technical expertise needed to produce better early warning systems, sophisticated new guided missiles, advanced aircraft, and a host of other weapons and defense measures was concentrated in such areas as Cambridge.

In the years following the war, scientists and engineers from a number of fields whose talents had been honed by working for OSRD, or for the myriad government agencies that emerged to carry on its work, began to take their expertise to the marketplace. With the growing distaste of many campuses for projects that lacked academic flavor, technical people found themselves becoming entrepreneurs either to pursue specific government projects or just to continue doing the kind of work they knew best. As a result, there arose in the region a new class of company that focused more on bringing research skills and high-tech expertise to the marketplace than on coming up with specific commercial products.

During the 1950s, Massachusetts firms received more than $6 billion worth of prime contracts from the DOD. The level of funding rose to more than $1 billion annually through the 1960s. Nonetheless, although the federal government might have provided an initial market for goods and services, its patronage by no means guaranteed success. Like small companies catering to commercial markets, many of the spin-offs did not survive. Out of fifty companies that had spun off from Lincoln Lab in the post–World War II period through the early 1960s, for example, twelve were already out of business when studied in 1965, and by 1988, only six of the fifty were still listed as independent companies in the *Directory of Massachusetts High Technology Companies* published by Mass Tech Times. Some of the firms merged with other companies, were acquired, or changed their names, but this figure still gives some indication of the high failure rate common among small firms.

Many of these university spin-offs were able to go into business only because of the availability of R&D contracts from the various new governmental agencies that sprang up to carry on work begun by OSRD, which was disbanded at the end of the war. The Armed Forces, resoundingly convinced of the value of the "basic" research the Office had conducted during the war, found themselves scrambling to refocus and restructure their own research activities.

This shift toward basic research led to a fundamental and far-reaching expansion of the federal government's role in supporting R&D. As the war drew to a close, for example, the Navy laid plans for a central agency to coordinate and increase its research and engineering development activities. On August 1, 1946, Congress voted to establish the Office of Naval Research (ONR), not only approving the Navy's involvement in R&D, but also formally acknowledging the role of the legislative branch in appropriating the necessary funds.

ONR quickly became a major force in shaping the patterns of research support in the United States. By 1948, it coordinated projects valued at about $43 million, nearly 40 percent of all government-funded research at the time. This figure was well below the Navy's peak R&D investment of $150 million during the war, but it was more than three times the amount spent in 1940. ONR not only employed a staff of 1,000 scientists, it also supervised contracts for over 1,100 projects at 200 other institutions, including universities. One of the crucial projects that ONR assumed responsibility for, however reluctantly, was Forrester's Whirlwind.

Similarly, late in 1945, representatives of the Armed Forces met to split up the radar research that OSRD had set up at MIT's Radiation Lab and Harvard's Radio Research Lab. The Army Signal Corps agreed to continue funding basic research at a new on-campus lab at MIT, the Research Laboratory for Electronics, created to carry on the microwave work begun during the war. The Army Air Force, which had already begun to recruit people from the Rad Lab, prepared to set up a new facility just off the northern edge of the MIT campus. The Air Force, made an independent branch of the armed services in 1947, came away from the Rad Lab breakup with fifteen projects and several tons of equipment, and by the following summer, the Air Force Cambridge Research Laboratories (AFCRL), as the new center later came to be known, had a staff of more than 1,100.

The AFCRL—with its three major research divisions, electronics, geophysics, and atomic warfare—played a leading role in the postwar development of radar technology and air defense systems. The center also initiated a new era of peacetime cooperation, particularly in the Cambridge area, between universities, the military, and industry. It became the cornerstone of a powerful military-oriented research infrastructure that grew up in the region, not only drawing in billions of dollars in defense contracts, but also helping to bolster small business by subcontracting R&D projects to a wide range of start-up firms. Because of the focus of its work, AFCRL had a close relationship with the Air Force's Project Lincoln, which ultimately became MIT's Lincoln

Laboratory. Although the two labs remained separate organizations, they eventually moved into neighboring buildings at Hanscom Air Force Base in 1954.

Four years later, this growing complex of research organizations spawned yet another key player. It had become clear to experts at Lincoln and in the Air Force that the SAGE air defense system, using a computer patterned on Forrester's Whirlwind, was not fast enough to deal with yet another generation of high-speed airplanes and missiles. One way to speed up the system was to integrate and automate the entire command and control process, from detection, through computer analysis, to direct communication with intercepting aircraft or missile sites. But the weapons systems available or under development all had their own unique specifications. Any attempt to integrate them would require detailed knowledge of these specifications. The industrial contractors of the different systems showed little interest in helping out. To make matters worse, the development team at Lincoln was the only group with the expertise and experience to oversee such an integration, and MIT—struggling to regain its academic independence and to distance itself from military matters—balked at the idea of getting involved further in the design of a refined defense system.

After considerable debate, MIT and the Air Force struck a compromise. In July 1958, the Institute spun off MITRE, a new, independent nonprofit company, to complete the project at hand and to undertake other engineering projects for the military. The firm was patterned after such nonprofit organizations as the Rand Corporation, a 1948 Douglas Aircraft Company spin-off, and the Sandia Corporation, which spun off from Western Electric in 1949. With the birth of MITRE, eastern Massachusetts was firmly established as a major center for electronics research and systems engineering, fulfilling Ridenour's prediction to Killian just a few years before.

One further development assured that it would remain so. In April 1961, the Air Force set up the Electronic Systems Division (ESD) at Hanscom, an organization that has evolved into the Air Force's chief center of expertise in command, control, communications, and intelligence technology—expertise that lies at the heart of modern warfare. ESD does no research and development itself, but it oversees the design, production, and deployment of systems for the Air Force, and is the chief contracting agency for MITRE and Lincoln Laboratory. ESD's 1990 budget of more than $3 billion, when compared with corporate sales figures, made it the third largest industrial organization in the state, surpassed only by Digital Equipment and Raytheon.

Although MIT has played a crucial role in the development of the regional and national military–academic–industrial complex, it has done so with some reluctance. MIT's leaders were anxious to get back to the business of education and research at the end of the war, but they found severing ties with the military difficult. The school's focus on advanced technology and its inherently practical bent, coupled with the considerable momentum built up during the war years, continued to draw military research funds to its laboratories.

Moreover, MIT researchers themselves had had a taste of the good life. During the war, the scale of research in terms of funds and equipment had reached heights hitherto unimaginable to the academic world. Flush with the OSRD experience, MIT's leaders at first saw no reason why research could not be funded by the military, as long as the work complemented its own educational goals.

One project area in particular, the Instrumentation Lab,

benefited from this pragmatic approach. Headed by the charismatic, inventive, and entrepreneurial Charles Stark Draper, a professor of aeronautical engineering, the lab made its reputation with the design and implementation of the Mark-14 gunsight. This electromechanical "black box," designed by Draper and his colleagues—including Jay Forrester—and produced by Sperry Gyroscope Company, used sophisticated gyroscopic control techniques to dramatically boost the accuracy of Navy antiaircraft guns. In 1942, in its first taste of battle, the device allowed the U.S.S. *South Dakota* to destroy all thirty-two of the much feared Japanese Kamikaze aircraft that attacked the ship.

The need for gunsights declined after the war, but demand rose steadily for accurate air navigation for long-range bombing missions over unfamiliar territory. Seizing the opportunity to further his research, Draper proposed to his contacts in the Air Force that the gyroscopic principles used so successfully to deliver gunfire to its target could be adapted to produce an airborne "inertial guidance" system, one that would require no external reference points to guide a craft from any point on the globe to another. The Instrumentation Lab took on its first contract to create such a novel system in 1945. In 1953, with his characteristic flair for the dramatic, Draper arranged for a B-29 guided solely by his latest system to fly from Boston to Los Angeles, where a conference on the feasibility of pure inertial guidance systems was being held. After a twelve-hour flight using no landmarks or radio beacons to help out, the craft was right on target, and Draper presented data gathered during the trip as proof of the technique.

In the new era of nuclear submarines that could stay underwater for months at a time, and missiles that could cover thousands of miles, extremely accurate guidance systems were essential, and R&D contracts flooded into the Instrumentation

Lab. Working closely with the military and such industrial contractors as Lockheed and Northrop, Draper's team of students and engineers designed and developed guidance systems for such missiles as the Titan ICBM, Polaris, Trident, and a host of other craft. In 1961, the Lab itself became a prime contractor for the design and development of the system that ultimately guided Apollo 14 all the way to the surface of the moon in 1969. During that year, the Lab's $54.6 million budget represented one-fourth of MIT's total budget. Draper had always insisted on sticking with engineering projects from conception to implementation, but the growing scale of his commitments to both industry and the military were making top MIT administrators uneasy. Many felt that such links compromised the school's independence and integrity as an educational institution.

By the early 1970s, the future of the Lab had become a hotly debated and divisive issue on campus. Protesters demanded that the Lab's considerable resources be redirected to peaceful uses. At the other extreme, supporters felt that the Lab should continue as part of the Institute because of the practical educational opportunities it provided for students. The final compromise pleased almost no one. On July 1, 1973, the organization became the Charles Stark Draper Laboratory Incorporated, an independent, nonprofit R&D company. Freed from the limitations imposed by MIT, the company's contracts ballooned to nearly $90 million in 1974, and climbed to more than $350 million by the end of the 1980s. By that time, more than fifty-five companies had spun off from the Laboratory.

It is not surprising that MIT's labs became a major source of spin-off companies. They provided practical expertise in a broad range of strategically important technologies, and the government offered a huge new market for the application of that expertise. Being able to claim an association with MIT was no doubt a plus for the budding entrepreneurs and starting a company to take on government R&D contracts was a way to continue doing top-notch research in a university-like environment. According to Howard Whitman, one of the founders of Dynamics Research, proposing and managing research projects for his own company was not much different from doing it back at the Instrumentation Lab. "It's not a bad business," Whitman asserts. "And you don't need a lot of capital to do R&D."[3]

## SCIENCE FOR PEACE

The extraordinary military-oriented R&D complex in the Route 128 region—consisting of small businesses, large defense contractors, universities, and government labs—owes its existence largely to the success of OSRD during World War II. In the decades since the war, defense spending in the area has spurred development in such diverse fields as microwaves, electronics, computers, guidance systems, and artificial intelligence.

These areas represent just a portion of the total federally backed R&D activity in the region, however. A great deal of research—and commercialization of new technologies—has taken place in such nonmilitary fields as energy, medicine, chemistry, and biotechnology. Although the government agencies backing this work have no ties to defense, most of them arose directly from Vannevar Bush's wartime creation of OSRD.

On November 17, 1944, when it was clear that the war was

winding down, President Roosevelt asked Vannevar Bush for recommendations on how to apply OSRD's wartime successes to peacetime. Roosevelt's letter conveyed a sense of hope and expectancy: "New frontiers of the mind are before us, and if they are pioneered with the same vision, boldness, and drive with which we have waged this war we can create a fuller and more fruitful employment and a fuller and more fruitful life."[4]

Bush, who later recalled he had been expecting such a letter, saw this request as an extraordinary opportunity to further the cause of basic research and the goals of higher education. He immediately assembled four committees of leaders from the scientific, medical, and academic communities to advise him on an appropriate path to take. He bound the reports together and summarized them in a volume entitled *Science: The Endless Frontier*. The full report was delivered on July 5, 1945, to Harry Truman, who had assumed the presidency when Roosevelt was felled by a stroke.

Bush's landmark report recommended the creation of what is now known as the National Science Foundation. Bush explained that the contributions of the scientific community during World War II arose largely from applying advances in basic science made over several previous decades. He insisted that the ideas and insights that had arisen from basic research were essential not only to military strength, but also to industrial strength. "Basic research is performed without thought of practical ends," he wrote. "It results in general knowledge and an understanding of nature and its laws. This general knowledge provides the means of answering a large number of practical problems, though it may not give a complete specific answer to any one of them. The scientist doing basic research may not be at all interested in the practical applications of his work, yet the further progress of industrial development would eventually stagnate if basic scientific research were long neglected."[5]

Legislation for creating a science foundation was introduced on the same day the report was released, but a lengthy congressional debate on the organization's administrative structure stalled its passage for five years. In the hope of keeping the agency insulated from politics, Bush had recommended that its director be appointed by its board. Truman wanted the President to control the appointment, however, and vetoed a bill passed by Congress in 1947 on those grounds. A compromise was finally reached in which all twenty-four members of a National Science Board and its director were to be appointed by the President with the advice and consent of Congress. Truman signed the bill on May 10, 1950.

While Congress was debating the concept of the National Science Foundation, other wheels were already in motion that would permanently cement the government's role as supporter of a wide range of basic research. On August 1, 1946, Congress voted to establish both the Office of Naval Research and the Atomic Energy Commission. Soon after, the legislative body also expanded funding for the National Institutes of Health. These agencies set the precedent for the creation of the National Aeronautics and Space Administration (NASA) in 1958. Today, agencies as diverse as the Department of Transportation and the Department of Energy support basic research as part of their mission.

The NSF and the government's other nonmilitary agencies, like the DOD, have often backed projects with eventual commercial potential. Perhaps the most dramatic example of this is the fundamental investigation into the molecular basis of life begun in the 1940s and supported through later decades primarily by the NSF and the NIH. By the mid-1970s, biologists, working principally in Massachusetts and California, had come up with the basic techniques of genetic engineering. This development laid the foundation for the frenzy of scientific and commercial activity in the 1980s in the still-emerging field of biotechnology.

## A CONTROVERSIAL LEGACY

In a sense, Massachusetts owes much of its good fortune to microwaves. Because of its position of leadership in exploring this esoteric technology prior to the war—both at MIT and at Raytheon—the state was a natural place in which to locate a large radar research project. The funds and people that flooded into the area changed the region forever. MIT was able to demonstrate convincingly the strategic importance of advanced technology, and became a favorite site for the government-sponsored research that blossomed after the war. Raytheon, for its part, became a major defense contractor, employing thousands in Massachusetts alone.

Moreover, microwaves proved the salvation of Forrester's computer project when he was able to convince the Air Force that his machine could provide the electronic brain to control a sophisticated radar defense network. The merger of these technologies led to a tremendous expansion of the Massachusetts research infrastructure, including the spinning off of Lincoln Laboratory and the nonprofit MITRE Corporation. This expansion, in turn, indirectly sparked the founding of Digital Equipment, which became the state's largest employer. This extraordinary chain of events, catalyzed by the region's culture of entrepreneurship, greatly accelerated the rate of high-tech start-ups in the area.

The links that arose from the war effort have not been without their critics, however. There has been an ongoing debate, for example, about the wisdom of direct ties between academia and the defense establishment. Many academics have charged that defense programs, because of their sheer size, can exert undue influence on the course of university inquiry, limiting the funds available to fields without a clear defense orientation. Vannevar Bush himself warned that "applied research invariably drives out pure," and that it is basic, undirected research that "creates the fund from which the practical applications must be drawn."[6]

The founding of the National Science Foundation was intended to help preserve the integrity of basic research. Indeed, the programs of this and other nondefense agencies have greatly broadened the scope of basic university research to include such promising new fields as biotechnology. Even so, some critics in industry object to the concept of government-backed research, regardless of its focus, arguing that because most university research is funded by the government, academia is not sufficiently focused on the needs of industry. The system, some claim, turns out students trained to work as academics rather than as industrial researchers, though industry employs the vast majority of technical graduates.

A final area of debate focuses on the real economic impact of federal spending—particularly defense spending—on the Massachusetts economy. The highly cyclical nature of government business can be damaging to companies that do no commercial work, for example. In addition, some critics charge that the defense and space markets are "artificial," drawing the resources of business away from the industrial and consumer markets that will, in the long run, determine their ability to compete in the global marketplace.

There are no easy answers to any of these challenges. It is clear, though, that World War II permanently changed the relationship among academia, industry, and the federal government, and created an extraordinary new array of opportunities as well as tensions. It ushered in a dramatic new era of growth and exploration that laid the groundwork for what some would call a "Miracle."

# 4
# The Road to the "Miracle"

THROUGHOUT THE FIRST HALF OF THIS CENTURY, MIT, Harvard, and the Boston area's other universities were attracting new blood to the area and producing entrepreneurs with innovative schemes. The companies they founded were too few, however, and their sites too scattered to have created any recognizable high-tech enclaves. More visible was the slow decay of the region's established industries.

This situation changed abruptly with the end of World War II. Hundreds of scientists and researchers who had been drawn to the area's academic institutions as part of the national defense effort now turned their attention from government projects to commercial ventures. Moreover, startling advances in areas such as microwave and electronics technology made it possible—and compelling—for new companies to start up. The area's first high-technology explosion had begun. M/A-Com, Wang Laboratories, Millipore, Thermo Electron, Itek, and Digital Equipment are but a few of the ground-breaking companies founded in the decade after World War II. At the beginning of this entrepreneurially fertile decade, as though on cue, workers completed the first section of Route 128.

The road's original charter was simple: to "provide ready access to the North and South Shore recreational and residential areas for traffic from the Metropolitan area and Western section of the state."[1] Historic Boston, with its narrow, winding streets laid out to meet the needs of a time now three centuries past, was proving to be a barrier to the growing volume of traffic seeking merely to pass through. The proposed Route 128 was designed to replace a narrow road by the same name that was closer to Boston and zigzagged through the hearts of nearby suburban towns.

The new Route 128 became the first limited-access circumferential highway in the country. It curved gracefully through miles of open land punctuated by pig farms, gravel pits, and vegetable fields. Despite its later fame, the highway got off to a decidedly inauspicious start. When the first 27-mile stretch opened in 1951, some critics dubbed it "the road to nowhere," referring to the highway's untenanted sweep. Such criticisms proved short-lived, though. In terms of industrial development, and the evolution of the state's high-tech community, the route's opening couldn't have been better timed.

Not only were a rash of new companies starting up in the post–World War II boom, but older companies looking for space to expand were finding the stock of existing buildings in Boston and Cambridge inadequate. For twenty years, during the Depression of the 1930s, and the war effort of the 1940s, almost no new industrial buildings had been built. The opening of Route 128 seemed like tossing candy to a pack of hungry kids. The highway suddenly exposed acres and acres of land a mere 15- to 20-minute commute to the all-important brain center of MIT and Harvard.

Much of the credit for the early development of Route 128 goes to Gerald W. Blakeley, Jr., who joined Boston's venerable

Cabot, Cabot & Forbes real estate management firm in 1948 expressly to try out a pioneering idea: the creation of a planned industrial park, providing tenants with a complete package, from development and commercial approvals, to financing, construction, and management.

One early settler along the Route 128 corridor—and one of the most influential to the region's continuing development—was the federal government. In 1953, the Air Force took over legal ownership of Hanscom Field, which was to be the home of its Cambridge Research Center; Lincoln Laboratory, operated by MIT for the Air Force; and the Air Force's Electronic Systems Division.

By 1955, just four years after its opening, Route 128 had begun to win fame as a sort of high-tech magnet. *Business Week* magazine that year ran a feature entitled "New England Highway Upsets Old Way of Life," referring to Route 128 as "the Magic Semicircle."[2] There were at least 99 companies employing 17,000 workers along Route 128 by 1957, according to an economic impact study that MIT prepared for the state, and 96 percent of the companies had moved out from within four and a half miles of the center of Boston. CBS Hytron, Sylvania, M/A-Com, Millipore, MITRE, Thermo Electron, and Wang Laboratories were just a few of the new arrivals. "By 1960, Route 128 was the place to be," asserts George Hatsopoulos, founder and CEO of Thermo Electron.[3]

Former Massachusetts Governor John Volpe, who served as Commissioner of the Department of Public Works from 1953 to 1956, recalls that the steady transformation that took place through the 1950s was startling. "The idea that it would develop into a high-tech highway was not the original purpose for building the highway, but as soon as it began to develop, it was clear that this would be an opportunity," he says. "Route

128 was the foundation stone for the development of high technology as we know it today. You couldn't possibly have put the number of high-tech firms that located along 128 into existing communities."[4]

In fact, the "road to nowhere" had become so congested that in 1958, a mere seven years after its opening, workers began widening the most heavily traveled stretch of the highway from six lanes to eight. Douglas Ross, founder and chairman of Waltham-based SofTech, still bemoans the loss of the fifty overpasses that had to be replaced. "Every bridge was cut stone," he recalls wistfully. "Every single bridge from end to end was a work of art. But by the time they finished the road, they started to tear them out to widen it. They didn't know it was going to spawn all this."[5]

By the end of the 1970s, the highway's popularity began to backfire. Space in prime locations had run out. Moreover, the daily volume of cars, which had already forced the road's widening, had completely eclipsed all expectations. By the late 1980s, traffic along heavily developed sections of Route 128—such as Burlington and Waltham, whose zoning laws encouraged industrial development—topped 150,000 cars a day, and was still increasing. "I wouldn't start here today—no way," grumbles Digital Equipment's Ken Olsen, whose Maynard headquarters lies ten miles west of Route 128. "I wouldn't ask people to drive these roads."[6]

Route 128's congestion, however, did nothing to slow the high-tech community's momentum. Many companies started up on or moved out to Interstate Route 495, which describes a second, larger arc, mimicking Route 128's curve, at a distance of about twenty-five miles from Boston. Others moved even farther west to Worcester, across the northern border into New Hampshire, or back in to areas like Cambridge's Kendall

Square. The "Route 128 Phenomenon" was turning into the "Massachusetts Miracle."

## THE MONEY TRAIL

Boston and Cambridge clearly had more than their share of talented innovators. But the brightest entrepreneurs with the best ideas would have gone nowhere without sufficient capital to back them. By the same token, Boston's investment community would not have roared into action had there not been the engine of new technology companies to drive it.

Nonetheless, the quest for capital by Boston-area entrepreneurs has been buffeted by cyclical ups and downs. The amount of money a company needs to get started has varied enormously with the year, the technology, the intended market, and even the ambitions of the entrepreneur. Similarly, the enthusiasm of investors has blossomed or withered depending on everything from stock market performance and federal tax policy to the popularity of certain areas of technology and the overall health of high-tech businesses.

For most entrepreneurs, one of the first hurdles after coming up with an idea for a new business is to raise venture capital—the money that gets a company on its feet. In exchange for cash, the investor typically receives stock in the company which—if the company thrives—can someday be worth many times the original investment when the company goes public or is acquired. Since no one can guess how successful any venture will be, or even if it will be successful at all, start-up companies are a risky investment, but the potential gains in new and largely unexploited areas of technology can far outperform other more secure investments.

Venture capital firms as we know them today began with the founding of American Research and Development Corporation (AR&D) in Boston in 1946, bringing to life MIT President Karl Taylor Compton's earlier plan, which had been put on hold during the war. AR&D had a novel and seemingly altruistic mission. Its goal was not only to make money, but to spur the creation of new companies in New England, and to bring important ideas into the marketplace. The innovative idea was not easy for the conservative Boston financial community to swallow, and the firm was just barely able to raise the $3 million it had set as a minimum for its own start-up. Early investors included Massachusetts Investors Trust, the John Hancock Mutual Life Insurance Company, and MIT and a handful of other universities. In all, the firm's investors eventually numbered more than 1,200.

French-born brigadier general and Harvard Business School Professor Georges F. Doriot was recruited to head the firm. Doriot, who died in 1987 at the age of eighty-seven, was the soul of AR&D; the pattern that he set still molds the venture capital community in Boston and that of the entire nation. Strong-willed, demanding, and uncompromising, Doriot made a lasting impression on all he met, whether lecturing about management skills at Harvard, where he taught for forty years, or counseling and cajoling his "family" of entrepreneurs. His reign at AR&D lasted twenty-six years, until Textron bought the investment firm in 1972. "Most successful people are controversial," muses William Congleton, a Doriot protégé at Harvard whom the general recruited to join AR&D in 1952. "Doriot was not an easy man to be around. Working with him was a trial. He was very demanding, very egotistical, very self-centered. But he was also charming and charismatic. It was very much Doriot's show because we all had so much respect and admiration for him."[7]

AR&D was also very much a creation of the elite. In addition to Merrill Griswold, chairman of Massachusetts Investors Trust, and Ralph E. Flanders, president of the Federal Reserve Bank of Boston, AR&D's board of directors featured such heavyweights as the president of Monsanto Chemical Company, William Rand; MIT treasurer Horace Ford; and John Hancock President Paul Clark. Many members of Doriot's top management came from the Harvard Business School, and the firm's three advisory board members were MIT President Karl Compton and two MIT professors.

True to its founders' vision, AR&D began by putting a premium on backing "noble" ideas—a concept alien to today's more competitive and profit-minded venture capital industry. Denis M. Robinson, who co-founded High Voltage Engineering Corporation along with MIT professors Robert Van de Graaff and John Trump, says that his company's initial mission to develop powerful X-ray machines for cancer therapy convinced AR&D to pick High Voltage as its second investment. "[MIT President] Compton's advice to Doriot was, 'You should put this Van de Graaff/Trump company in your portfolio,' " he recalls. " 'They probably won't ever make any money, but the ethics of the thing and the human qualities of treating cancer with X rays are so outstanding that I'm sure it should be in your portfolio.' "[8] AR&D invested $200,000 in High Voltage, and, despite Compton's pessimistic prediction, the investment was worth $1.8 million when High Voltage went public eight years later.

By 1953, just seven years after its founding, AR&D had been flooded with over 2,300 proposals. It had invested some $4 million of its own money and another $1.5 million of co-investors' money in sixteen major ventures and four very small ones, and the sales of companies included in the portfolio had reached $60 million. Of the sixteen larger companies, half were pre–World War II manufacturers of well-established products, such as spectrographs and timing devices. The other half, founded since the beginning of the war, were primarily based on "new" technologies, including radar systems, automatic control systems, and rocket motors. Ironically, only half the ventures were based in New England, with some located as far away as Texas.

Doriot's emphasis on choosing the person rather than the product most characterized AR&D's investment strategy, and still influences many venture firms today. His oft-quoted philosophy—better a brilliant man with a mediocre idea than a mediocre man with a brilliant idea—governed his investment choices. Once the right entrepreneur came under AR&D's wing, Doriot liked to serve as a mentor, offering management advice and strategic direction in addition to cash. Not surprisingly, the intensity of Doriot's relationships with the companies he sometimes referred to as his "children" resulted in some fallouts with entrepreneurs.

Despite Doriot's drive and enthusiasm, the early years at AR&D weren't easy. Although the skeptics who shook their heads when AR&D was formed never got the pleasure of seeing it fail, neither could the venture be counted a quick success. "Nobody had ever done that kind of thing before and all of us were feeling our way," recalls Congleton. "What is an entrepreneur? Who do you choose? How do you work with them when you've made an investment? Those were the questions we were trying to answer."[9] As Doriot always liked to stress, AR&D was not in the business for short-term gains.

The firm did not operate in the black until 1951, five years after its founding, and then with a profit of only $46,000. But it stuck to its original vision. "It takes a long time to build up companies," AR&D's annual report noted that year. "It is hoped that lack of return on investments today will mean greater values later."

## THE FRUGAL FIFTIES

A glowing article on AR&D in the November 1952 issue of *Fortune* magazine won the firm national attention, but there was no rush to follow in the pioneering company's footsteps. Other than a few family investment groups and professional trustees, AR&D had the Boston area largely to itself. Similarly, in New York and California there were only a handful of firms, including the newly formed J. H. Whitney & Company, the Rockefeller organization, and Draper, Gaither & Anderson, a Palo Alto firm that eventually folded in the mid-1960s.

"There was no over-the-counter market for new issues of stock really until the early 1960s," explains Peter Brooke, the ebullient founder of two Boston venture capital operations, TA Associates and Advent International. "So there was no clear-cut indication that investing in these companies would lead to handsome returns."[10] As a result, most of the high-tech firms that began to build and grow along newly opened Route 128 in the early 1950s faced the same financing options as the startups of the previous few decades.

Although the established venture capital industry subsequently entered the limelight—with its partnerships of highly specialized professionals investing billions of dollars of mostly institutional funds—that was just a piece of the investment picture. The seed and early round money that gives a start-up its first chance to set up an operation and begin production comes from a diverse array of sources, including not only professional venture capitalists, but also friends and relatives of entrepreneurs, government agencies, self-made businessmen, corporations in search of new technologies, and, in many cases, the entrepreneurs themselves.

While it was not unusual for a company in the 1980s to raise anywhere from $2 million to $10 million in venture capital in its first year, launching a company in the 1950s did not require such massive infusions of cash. Part of the discrepancy between the amount of venture capital invested in the 1950s and 1980s can be explained by inflation, but there were other reasons why companies required less capital then. For one, there had been very few broadly publicized "overnight success" stories, and both entrepreneurs and investors were more modest and more patient. Most had no expectation of creating a major new corporation, and in fact, few did. The high-tech community largely comprised small consulting firms, contract R&D outlets, and specialty hardware producers, selling their high-tech expertise. The founders of many of these small, but essential, companies were driven mainly by the desire to be independent and to do what they liked to do best. With no big-time investors breathing down their necks, they felt little pressure to impress the marketplace within the first few months of operation.

When An Wang set up Wang Laboratories as a sole proprietorship in 1951, for example, he did so at the age of thirty-one, fresh out of Harvard's Computation Laboratory, with nothing to his name but $600 in savings and a then-disputed patent for magnetic core memory. "My expectations were very different from those of many of the start-ups in high technology today," Wang, who died in 1990, wrote in his autobiography, *Lessons.*

"I did not anticipate becoming wealthy. Nor did I have a chart showing steeply curved earnings projections or a business plan. And I did not have the worry of risking somebody else's money."[11]

Wang's business at first consisted of making and selling the memory cores he had developed at Harvard. Only gradually did the company move into desk-top calculators, and then computers. It was not until 1976 that Wang introduced the first of the word processing computers that became the lifeblood of the company. Wang's annual sales soared to almost $3 billion in 1988 before beginning a downward spiral the next year. Wang never did seek venture capital, although in 1959, facing a cash crunch, he sold stock worth 25 percent of Wang Laboratories to a Cleveland-based machine tool company, Warner & Swasey Company, in exchange for $50,000 and access to $100,000 in loans.

The founding of Dynamics Research Corporation in 1955 was even more informal. Howard Whitman, one of six cofounders, says he abruptly quit his job at MIT's Instrumentation Laboratory after completing an ongoing project. "It was a good opportunity for a break," he explains. "Four of us from the Lab brought in a patent attorney and a friend as participants and we started the company in the dining room of a two-family house. Each of us chipped in $100."[12] The new firm soon got its first job: a $6,000 subcontract from RCA to do work on an airborne fire control system. By 1957, when the Polaris missile program began, Dynamics Research had been able to carve out a solid niche in the defense business. Not until the early 1960s did the company raise professional venture capital, and it went public in 1965 to allow some of the original partners to cash in their stock.

Millipore Corporation, a maker of membranes and other

products for analyzing and purifying liquids, started off in 1954 without professional venture capital financing, but not for want of trying. Founder Jack Bush complains that in his initial search for money, being the son of renowned MIT Professor Vannevar Bush actually worked against him: Local investors were afraid that if the deal somehow went bad, it might hurt their relationship with the influential and respected senior Bush. Ultimately, however, his father's connections, in particular his secretary and assistant at MIT, Carroll Wilson, helped grease the wheels. Through Wilson's introductions to potential investors, Bush was able to raise $300,000, primarily in individual contributions ranging from $5,000 to $15,000 each.

In part because of his own discouraging experience, Bush believes that professional venture capital is only a limited solution to the financing needs of young companies. "I'm not sure that venture capital companies in general are very effective in supporting the start-up ventures," he declares. "They've got to see documentation, ad nauseam, when you know some of these things ought to be played on the basis of, 'Let's give it a whack! If it goes, great. If it doesn't, we've lost our dough.' " He adds: "The venture capital company in general is ultra conservative. They've got to defend their decisions if something goes sour. So taking a kid like myself who had no business experience, in a field where there were no parameters to measure, with a volume of sales that wouldn't support the enterprise, they couldn't do it. But on the other hand, the individuals who took a ride on this did very well."[13]

Although Route 128 and Boston's high-tech community were just getting established in the early 1950s, outside corporations had already begun to turn to Massachusetts to buy into what they saw as exciting new technologies complementary to their own businesses. In 1952, Dana W. Atchley, a Harvard graduate with a broad range of experience including stints at Sylvania, Tracerlab, and MIT's Radiation Laboratory, was hired by United Paramount Theatres—later to merge with ABC—to find a new business worthy of investment. Atchley, for his part, wanted to run a company.

His first choice was Hytron, a local radio tube company, but CBS bought it first. "When CBS went and bought it, that gave me credibility that I did not deserve," Atchley grins. "I think CBS lost $9 million before they dumped it. But at the time I looked very credible."[14] Instead of Hytron, Atchley chose M/A-Com, then known as Microwave Associates, a 1950 Sylvania Electric Products spin-off that was strapped for cash. Although the four founding engineers had quickly landed lucrative government and commercial contracts to supply such sophisticated components as magnetrons—the power generators for radar and other microwave systems that had emerged from the war effort—they were unable to raise enough money to finance M/A-Com's rapid growth, even after borrowing as much as they could from friends and families. United Paramount paid $125,000 for 50 percent of M/A-Com, which at the time had only eight employees and sales of $80,000.

"They were desperate for financing," Atchley recalls. "In those days you couldn't walk up and down the street and just get it." Although M/A-Com prospered, Atchley claims he encountered some Brahmin resistance the next time the company needed funds. Before M/A-Com went public in 1957, Atchley turned to Lehman Brothers in New York after finding the doors of the Boston financial community closed to him. "The old New England types didn't like the fact that we had two blacks, and a Greek, and one tired Harvard guy," he contends. "They would have liked a little fancier social crowd, whereas the New York crowd loved us. It was hard to find money locally, but easy in New York."

## IGNITION

Although investment money flowed slowly in the early 1950s, later that same decade several events had a profound impact on creating new financial options for high-tech companies. One of the most momentous was the 1957 Soviet launching of Sputnik, the world's first man-made satellite. It was the middle of the Cold War, and the United States reacted quickly to this Soviet coup, pumping money into military and space R&D efforts—including NASA, launched in October 1958—to allow America to catch up and take over the lead in the race into space. Because of Boston's growing concentration of high-tech activity and expertise, a good deal of this money poured into the Route 128 area.

In addition to multimillion-dollar R&D contracts for larger organizations such as Raytheon, EG&G, Lincoln Lab, and the Air Force's research facilities at Hanscom Field, an entire network of smaller companies was drawn into the mobilization effort through subcontracts. The rush of federal dollars was a boon for recent start-ups, as well as helping to spark the creation of companies like Geophysics Corporation of America—later dubbed GCA—an Air Force Cambridge Research Center spin-off.

The Bank of Boston, the area's largest bank at the time, responded aggressively to the surge of technology companies needing cash in the late 1950s. Then vice president and later chairman William Brown, along with TA Associate's Peter Brooke, who was then a loan officer at the bank, sought out prospects along Route 128, often lending against the value of a company's federal contracts and receivables. The unofficial wooing of this new sector won a slew of new customers for the bank, and also got much-needed capital out into the high-tech community. Wang Laboratories, Unitrode, Teradyne, Thermo Electron, High Voltage Engineering, and Analog Devices were but a few of the companies that eventually borrowed through the high-tech program. According to Analog Devices founder Ray Stata, it was a Bank of Boston loan that made it possible for him to forgo venture capital financing for the four years before he took his company public in 1969.

The growing high-tech sector turned out to be a gold mine for the bank. "In the early sixties we did a survey and figured we had accounts from 85 percent of the high-tech firms on Route 128," recalls Brown.[15] Adds Brooke: "That bank had more to do with the development of the high-tech industry here than any other capital source."[16] Driven by the Bank of Boston's success, other area banks began to beef up their lending programs to the high-tech industry.

During this same period, the federal government got into the act with its Small Business Investment Company program, signed into law by President Eisenhower in August 1958. The program, intended to help get capital to small businesses, allowed for the licensing of SBICs, private investment companies that could leverage their investment dollars by borrowing additional funds from the Small Business Administration (SBA). Many critics thought it was a flawed concept from the start. A number of participants set up SBICs just to get access to the federal loans. Even for the legitimate players, it could be dangerous to invest SBA money in start-ups. The money that an SBIC borrows from the government has to be repaid, but an investment in a start-up is a high-risk proposition that may not pay off for many years—if at all.

Nevertheless, the SBICs signaled a turning point for the venture capital industry, providing the legitimacy it had been trying to win. "The SBIC program really got the industry started," contends Jane K. Morris, vice president of Venture Economics, Incorporated, the research, consulting, and publications firm specializing in venture capital. "Many partners of the old guard came from SBICs."[17] In addition, by forming an SBIC, a bank could own stock and make venture capital investments for the first time since the early 1930s. Here again, the Bank of Boston was a leader, receiving the fourth SBIC license in the country, and the first in New England. Richard Farrell, who managed the bank's SBIC from 1961 until 1980, agrees that although SBICs may not have been a great concept overall, they worked well for banks. In the case of Bank of Boston, for example, the institution did not rely on borrowed money from the SBA when making long-term, high-risk investments, so it did not have to worry about getting its money out in time to repay federal loans. One of the bank's most successful deals was a series of investments totaling $150,000 in Damon Corporation in the early 1960s. By 1969, its investment in the electronic instrumentation and clinical testing company was worth about $10 million.

As the 1950s gave way to the 1960s, both entrepreneurs and venture capitalists finally got the role models they had been waiting for. In 1957, the same year Sputnik was launched, Digital Equipment Corporation (DEC) and Itek Corporation were founded, the first a spin-off from MIT's Lincoln Laboratory, and the second a spin-off from Boston University. DEC's success has become a local legend, and it was the making of American Research and Development, the venture capital firm that put its faith in founder Ken Olsen. Ten years after DEC's founding, and one year after the company went public in 1966, AR&D's $70,000 initial investment in Digital was worth more than $100 million, and Ken Olsen was a millionaire. By the time Textron, a Rhode Island–based conglomerate, bought AR&D in 1972, the DEC investment had grown to more than $400 million.

Itek brought similar fame to Laurance Rockefeller's venture group. According to Walter J. Levison, one of the founders, the small group of scientists and engineers from Eastman Kodak and Boston University who started Itek had planned to develop microfilm and microfiche products to store graphic information. Their charter soon shifted, however. Just a matter of weeks after the company's founding, the Soviets launched Sputnik, and Itek found its expertise in lenses in great demand. Over the next decade, the company became known for its tracking telescopes and aerial cameras used in reconnaissance work. Science stocks at the end of the 1950s were booming, and Itek was at the top of the heap. In two years, the company's stock went from $2 to $345 a share.

The experiences of DEC and Itek on the East Coast, along with the more loudly trumpeted successes of companies like Fairchild Semiconductor and Scientific Data Systems on the West Coast, helped fuel an entrepreneurial frenzy in the Boston area. Teradyne, Unitrode, Damon, Compugraphic, Bose, Analog Devices, Cullinet Software, Data General, and Computervision were among the hundreds of high-tech companies launched in the 1960s. Although the collapse of the stock market in 1962 temporarily dampened the prospects for taking companies public, a model for success had been established.

As new technologies and product ideas continued to pour out of the area's universities, government research labs, and companies, a new wave of venture capitalists rose to meet them. "It was a paradise here in the sixties," exults Peter Brooke.[18] Brooke left the Bank of Boston's lending program in 1961 to run Bessemer Securities, a family-funded venture firm in New York, but he was back in Boston by 1963 to establish an investment banking business for the brokerage firm of Tucker, Anthony & R. L. Day. These efforts were so successful—in five years his first $100,000 of investment had grown to $2.5 million—that in 1968, Brooke set up a venture capital subsidiary for Tucker Anthony named TA Associates. By 1978, Brooke and some colleagues had bought out Tucker Anthony's interest in the successful venture firm and were running it on their own.

## STALLED

Venture capitalists like TA Associates were driven by success stories from both coasts. But the market and investment successes of the mid- to late 1960s that inspired the venture capitalist expansion were not to last. The 1970s were lean years for investors.

In 1970, the United States slipped into a recession that jacked up nationwide unemployment to a temporary high of about 6 percent by 1971. In Massachusetts, unemployment kept climbing, reaching 8 percent by the beginning of 1972. More than fifty manufacturing plants in the state closed in 1971 alone, and more than 112,000 jobs in manufacturing industries were lost in the five years between 1967 and 1972—most from the aging remains of the once-vital textile, apparel, and leather goods industries. Then the recession of 1973—the worst since the great Depression of the 1930s—really knocked the state for a loop. In 1975, unemployment in Massachusetts peaked at 12.3 percent, considerably worse than the national average of 8.5 percent. Few people expected much help from the fledgling high-tech industry, which, by some estimates, accounted for only about 10 percent of employment at the close of the 1960s. Moreover, high tech didn't seem to be faring any better than the region's old-line industries.

Changes in federal programs were partly responsible. Once NASA achieved its goal, in 1969, of landing a man on the moon, the space program lost momentum and the government cut back on the R&D spending that had benefited many Boston-area start-ups. In addition, the turmoil surrounding the quagmire in Vietnam was dampening enthusiasm in Congress for sponsoring defense research and development. As a result, defense contracts awarded to New England firms and research institutions plummeted by almost 40 percent in real terms from 1967 to 1971; almost 15 percent of the manufacturing jobs lost during that period were in high tech. In the early 1970s, the state's best hope for the future seemed to hold little promise as laid-off scientists and engineers cast about desperately for work, and in some cases, left the region for good.

Tax policies didn't help the situation. In 1969 Congress hiked the capital gains tax from 25 percent to a whopping 49.2 percent, bringing it in line with the maximum rate on ordinary income, and lowering the incentive to make high-risk investments. The blow to risk-taking dealt by the tax law change was worsened by the nation's increasingly shaky economy. The recessions of 1970 and 1974 effectively dried up whatever enthusiasm was left for venture capital and new public offerings on the stock market. According to Venture Economics Incorporated, venture capital firms committed only $466 million of new capital between 1970 and 1977, and only $10 million in 1975. "The venture capital business reached its bottom in 1975," declares Richard Farrell, then at the Bank of Boston. "The record from 1969 to 1975 was terrible, and anybody that got in and out during that time frame did terribly."[19] Because the market for new public issues was effectively dead—one study showed that only four small high-tech companies nationwide made initial public offerings in 1974, compared with more than 200 in 1969—the only way for a company to raise capital was to merge with another company or find a buyer.

The crisis of the early 1970s threw the deterioration of the Massachusetts economy into dramatic focus. The old manufacturing base was crumbling, and the high-tech community, despite its touted potential, was apparently too dependent on federal funding to stand on its own. Not only were the corporate and individual tax bases shrinking—as both businesses and workers left the state—but state government spending was rising at an alarming and unsupportable rate. In his first term as governor in 1974, Michael Dukakis faced a $350 million state deficit that grew to $500 million the following year. In 1975, Governor Dukakis imposed a "temporary" 7.5 percent surcharge on state income taxes—not repealed until 1986—to help meet the state's burgeoning revenue needs.

In part to support the large numbers of unemployed, as well as to provide other state services, the tax burdens on both corporations and individuals had grown to rank among the nation's highest. On a per capita basis in fiscal year 1976, Massachusetts ranked third in the nation for state and local property taxes, fourth for state corporation income taxes, and fifth for state individual income taxes.

This was the period that gave rise to the infamous, long-lived, and recently revived title of "Taxachusetts," which became a rallying cry for the business community during the late 1970s and early 1980s. "At that time, the number one problem was attracting talented people to the state of Massachusetts," declared Ed de Castro, who had founded Data General in 1968, and who was forced out of the company at the end of 1990. "We couldn't get them to move into Massachusetts, and when you dug deeper, you found that among competing industrial states, Massachusetts was way off the deep end in terms of personal taxation."[20] De Castro was not alone. "As you were recruiting people, they would just say, 'I'm not going to work for you because the taxes are too high,' " claims Alex d'Arbeloff, founder and chairman of Teradyne. "The government almost ran the whole high-tech movement out of business in the mid- to late 1970s because of this taxing problem."[21]

Needless to say, many entrepreneurs recall this period grimly. Douglas T. Ross, founder of SofTech, remembers making a last-ditch attempt to raise start-up money in the fall of 1969 from the Boston office of a New York investment firm. "I went to them and they knew I was tapping the kiddies' college fund, which was literally true," he says. "They knew I was in dire straits, so they made me this really ugly offer—less money for more equity than anyone had even hinted at before. When word gets out on the street, in those days especially, it was a very closed community, so they thought they had a sure thing—the vultures!" Ross turned them down and was eventually able to raise a few hundred dollars from individual and institutional investors on his own, "enough to keep the wolf from the door."[22]

Bill Foster, who founded Stratus Computer in 1980, claims that what little venture capital there was almost all went to the flashier Silicon Valley area, which already had a higher profile than Route 128. "I had a couple of venture capitalists from New York tell me, 'You'll never raise any money for a computer start-up in New England,' " he recalls. "It was really a psychology that said California is where all the start-ups are and that's where all the action is. We were the first computer company to be started on the East Coast since Prime in 1972."[23]

Raising the money to start Prime was not an easy task, either. It took a year for the seven co-founders—all from Honeywell's computer division on Route 128—to find an investor. "We went to every rat hole we could think of," recalls founder Robert P. Berkowitz. "Very few people had a vision of what the computer industry was going to become." Berkowitz finally contacted David Dunn of Idanta Partners in San Diego. "The first thing Dunn said was, 'I have a motto. I never have and never will invest in a minicomputer company,' " grins Berkowitz.[24] Despite this inauspicious beginning, the founding group, which included Bill Poduska of Apollo and Stellar, convinced Dunn that their venture was worth a $300,000 initial investment. Dunn's involvement did not end there. He remained chairman of the board at Prime until the summer of 1989, when New York venture capital company J. H. Whitney & Company bought Prime in a friendly merger valued at $1.25 billion. Russell Planitzer, who orchestrated the merger, had been vice president of marketing development at Prime until 1981, when he left to join J. H. Whitney.

Although Prime's founders had greater cash resources than many entrepreneurs, Poduska says they never considered bootstrapping the company. "It's a very poor venture that's launched with personal money," he declares. "A guy who mortgages his house is just too worried about other things to run his business right."[25]

Raymond Kurzweil, prolific founder of three Boston-area companies, had an even harder time in 1974 launching his first company, Kurzweil Computer Products. The company's first product, the Kurzweil Reading Machine, was an innovative computerized aid for the blind that could scan printed text and read it out loud. The machine was lauded as a breakthrough, winning Kurzweil public service awards from Governor Dukakis and President Reagan, but it did not win him much funding. With no proven track record, and a dismal venture capital market, Kurzweil was able to raise only $33,000 in seed money from Johnson & Johnson's venture capital subsidiary to supplement the $100,000 he raised from family, friends, and other small investors in the company's first two years. "Raising money, holding off creditors, and keeping people's morale up was a big thing then," he recalls. "It was 1978 before we got our first significant venture capital."[26] That money, $930,000, came from Fidelity Venture Associates and Xerox Development Corporation, another corporate venture capital group. In 1980, Kurzweil sold the cash-strapped company to Xerox for $6 million.

## RECOVERY

Despite the prevailing views of prominent investors and entrepreneurs, the 1970s were not as unproductive as many came to believe. Mass High Tech's 1989–90 *Directory of Massachusetts High Technology Companies* lists only 351 high-tech companies founded during the "booming" 1960s, but there are 638 surviving companies born in the "depressed" 1970s. Although some of this discrepancy can be explained by the greater attrition rate that is inevitable when dealing with an earlier decade, these figures clearly indicate that entrepreneurial activity still thrived in the 1970s.

A closer look at the start-ups of the 1970s reveals that most were relatively small operations, often privately held, that could rely on their own resources. Companies that did consulting work or specialized in narrowly focused government applications often funded their minimal capital needs through contracts, just as many of the firms of the 1950s did. Roughly half of the companies were responding to the increased availability and accessibility of computers by developing new software programs, often targeting them at specific niches such as banking or data processing.

Moreover, by 1975 the nation had begun to climb out of its recession, and the Bay State was carried along for the ride. Defense research came back in favor under Ronald Reagan's administration, and the real value of contracts awarded to regional firms soared by 42 percent between 1976 and 1979. By 1980, Massachusetts, with just 2.6 percent of the nation's population, was capturing 5.6 percent of prime federal research contracts—contracts that were almost all won by the high-tech industry.

Demographic trends also aided the recovery. The growth of the labor force was held in check by the region's low birthrate as well as by the exodus of many workers to the Sunbelt. The population of Massachusetts stayed essentially stable through the 1970s, and increased only about 5 percent in the 1980s—half the national rate. And because Massachusetts added new jobs at the same brisk clip as the rest of the country, unemployment in the Commonwealth plunged relative to the nation overall. From 1975 to 1988, unemployment in Massachusetts tumbled from 11.2 percent—the highest in the nation—to a slim 2.7 percent—the lowest of the major industrialized states.

Although much of the state benefited from this recovery, there was little doubt that the most impressive job growth was coming from the high-tech sector and what gave this sector real economic muscle was its explosive expansion into the commercial marketplace. Leading the way were DEC, Data General, Prime, Wang, and the region's other minicomputer makers.

In the 1960s, the sophisticated and relatively costly machines made by these companies had sold mostly to a narrow range of federal agencies, and scientific and technical markets. But continuing advances had made the machines more flexible, easier to use, and a great deal less expensive. By the mid-1970s, as the nation emerged from its recession, a new generation of improved minicomputers had burst onto the marketplace. Driven by cheaper and more powerful hardware, a slew of software companies rose up to tailor specific computer applications to the needs of almost any business. As a broad range of industries embraced computer technology, the potential for both hardware and software companies seemed almost limitless.

The success of the computer business, coupled with a justifiable wariness of the often-fickle world of federal contracts, effectively reoriented the Massachusetts high-tech community toward the commercial marketplace. The share of the state's high-tech business being done for the government fell from 60 percent to just 25 percent from 1965 to 1980, though the dollar value and volume of these contracts rose substantially over the same period. Hundreds of firms in other areas such as artificial intelligence, robotics, and biotechnology sprang up to cash in on dreams of commercial success.

Route 128's heady rate of expansion continued, driven by the proliferation of new technology-based businesses. In 1965 there were 574 companies along the highway. Eight years later, that number had more than doubled to 1,212. Waltham, the most densely developed of the seven cities and twenty-three towns located along Route 128, was home to 186 separate companies and divisions, including branches of such out-of-state operations as Hewlett-Packard, Honeywell, IBM, and GTE, all hoping to benefit from the booming local high-tech infrastructure.

The Commonwealth of Massachusetts was a direct beneficiary of the boom. High-tech manufacturing jobs shot up by more than 47 percent from 1975 to 1983, pushing up total manufacturing jobs by almost 4 percent and more than compensating for the continuing decline in the still ailing non-high-tech manufacturing sector. Roughly 75 percent of the state's new high-tech manufacturing jobs were evenly divided between electronics and a category called "office machinery," which included computers. The greatest surge in jobs came not in manufacturing, however, where for all practical purposes, the high-tech sector was merely arresting the overall decline, but in the service sector. This sector grew by 46 percent between 1975 and 1983, and employed more people in the state by 1983 than any other job category.

Again, high tech was the driving force behind the growth. The business services category—which included such hot areas as software and data processing, as well as the more traditional advertising and public relations—grew 95 percent over the eight-year period, creating roughly a third of all new service jobs. Moreover, there was substantial growth in sectors indirectly related to high tech. The region's health services chipped in nearly 27 percent of all new service jobs, and education—a big business in a state with more than 120 colleges and universities—added almost 11 percent.

Although companies like DEC, Raytheon, and EG&G were among the most obvious contributors to the state's low unemployment, many of the new jobs were created by small businesses. According to MIT economist David Birch, the birth and expansion rates of small businesses in Massachusetts and the United States overall have been so great that these companies have been the key net producers of new jobs, despite a far greater failure rate than large firms.[27] Nationwide, companies with fewer than twenty employees create nearly 98 percent of net new jobs, Birch claims. In 1986, he says, 83 percent of all businesses employed fewer than twenty workers.

The entrepreneurial explosion of such small firms in the 1970s and 1980s, coupled with the growing success of more established high-tech concerns, dramatically demonstrated the potential of new technologies for bolstering a region's economy. Just as "Route 128" became the accepted term for describing the concentration of high-tech firms in eastern Massachusetts, "the Massachusetts Miracle" became the catchphrase for describing the impact of the region's entrepreneurial academic–medical–defense complex on the economy. The economic turnaround in Massachusetts in the late 1970s, and the period of growth and prosperity that followed, clinched the region's reputation as the economy of the future.

By the mid-1980s, there were almost 3,000 high-tech companies in the state. But these "new" industries did not pop up out of nowhere. They arose out of a firmly established high-technology community that had won attention as a potential economic force as far back as the 1940s. It was this sprawling web of companies, institutions, and attitudes that laid the groundwork for the "Miracle"; the economic turnaround of the 1970s was, in many ways, merely the tip of the proverbial iceberg.

## VENTURE CAPITAL HITS THE BIG TIME

The emergence of the high-tech community had a profound impact on venture capital. It ushered in a new golden era of venture investing. Several factors came together at once to push the industry forward. For starters, Congress finally responded to pressure from disgruntled venture capitalists and entrepreneurs in 1978, cutting the tax on capital gains back to 28 percent. In addition, the market for computers and software exploded as technological advances made computers more powerful, more affordable, and easier to use. The vast commercial potential of these machines drew scores of new entrepreneurs into the marketplace.

On top of this, as the nation's economy rebounded, the market for new issues came alive again in the late 1970s, rewarding those venture capitalists who had hung on to their investments through that dismal decade. Some eighty-one companies went public for the first time in 1979, raising a record-setting $506 million. "Those in the venture capital business in the late sixties and seventies were building up great value, but it just wasn't visible," explains Brooke, who had launched TA Associates in 1968. "By 1980 we were showing rates of return on a compounded annual basis of 30 to 40 percent. That's when a tremendous inflow of capital took place."[28] Specifically, private investors committed a stunning $13.4 billion in new capital in the seven years between 1978 and 1984.

Brooke refers to these years as "the wake-up period," and without a doubt, the high returns garnered by companies making their initial public offerings in the late 1970s caught the attention of both private and institutional investors. When the Department of Labor in 1978 cleared the way for a portion of pension fund assets to be invested in potentially higher risk—but higher return—ventures, the money poured in. This signaled a radical change in the makeup of venture capital investments, and by 1983, pension funds were responsible for almost one-third of the $4.5 billion in new capital committed.

Managers of university endowment funds, which weren't governed by the same laws as pension funds, also took action. Although Stanford had been investing a small portion of its endowment fund in venture activities since 1968, the more conservative New England universities had largely ignored the venture capital market, as well as opportunities to invest in spin-offs. Even MIT, despite its pivotal early involvement with American Research & Development, had sold its AR&D shares in 1955, and maintained a low profile thereafter. There was no law preventing the investment of endowment money in higher-risk securities, but the so-called "prudent man rule," used as a guideline since the 1800s, had led most endowment managers to conclude that venture capital investments would be too risky and not in line with how "men of prudence, discretion, and intelligence manage their own affairs."[29]

In addition, universities were generally wary about investing directly in the work of faculty members. Some university administrators believed that making such direct investments could compromise their institution's independence and academic integrity. Others feared that unless they invested in all faculty spin-offs, they might be charged with favoritism.

As the venture capital industry began to rack up impressive gains in the late 1970s, however, most of the major universities in the Boston area began putting part of their endowments in high-risk investments. The scope of each school's effort varied tremendously: Boston University led the pack with the 1975 launching of its Community Technology Foundation; MIT, ironically, remained a more cautious investor; but Harvard, with its hefty endowment and pension reserves, became one of the nation's most active universities in venture investing.

As endowment and pension funds flowed into the venture capital market, and the record of success stories mounted, raising money to invest in high-tech start-ups became almost as easy as hanging out a shingle. The sudden increase in the availability of money was like a shot of adrenaline for would-be entrepreneurs. Continuing advances in technologies with commercial potential fueled the fire. In 1980 alone, computer makers Stratus and Apollo, robotics company Automatix, and MIT Artificial Intelligence Lab spin-off Symbolics were all launched with initial venture capital investments ranging from $1.7 million to $5.3 million. The next year saw a wave of biotechnology start-ups, including Repligen, BioTechnica International, Integrated Genetics, and Genzyme. In 1982, Mitchell Kapor launched Lotus Development Corporation, setting in motion one of the personal computer software industry's most talked about success stories.

Lotus was on the fast track from the start. Kapor launched the firm with about $300,000 of his own money, earned on the royalties for programs he had developed for California-based software publisher Personal Software Incorporated. When that money ran out, Kapor won the backing of Sevin Rosen Management and Kleiner Perkins Caufield & Byers, who invested $1 million in the spring, and another $3.7 million in the fall of 1982. A good chunk of that money went toward an unprecedented marketing blitz: In an industry that had traditionally ignored marketing in favor of word-of-mouth publicity on ease of use and technical merits, Lotus spent more than

$1 million to promote its new 1-2-3 integrated business software before the product was even shipped.

Kapor's 1-2-3 may not have been the first integrated software package, but it was the most ambitiously marketed. Within two months of its introduction in January 1983, 1-2-3 was topping software best-seller lists nationwide. That same year, Lotus earned $14 million on sales of $53 million, a record for a start-up company. When Lotus went public in October 1983, a mere 21 months after its founding, the new shares were snapped up in minutes, adding $36 million to Lotus's corporate treasury. Kapor, who was all of thirty-two years old, sold 300,000 shares of his own stock for $5.4 million, hanging on to another 2.8 million shares with a value of more than $70 million.

Lotus's meteoric rise had a profound impact. It helped to inspire a surge of competitors, like Norwood-based Ovation Technologies Incorporated, in addition to companies targeting different niches of the software market, like Spinnaker Software, a maker of educational software. And it spurred venture capitalists nationwide to scramble to include personal software companies in their portfolios. It also set new standards for marketing and promotion, which in turn upped the cost of launching a software company. Before Lotus, many successful software firms had started up with little more than a personal computer and a telephone. After Lotus, a venture capital–backed marketing assault seemed almost a prerequisite for big success.

With venture capitalists competing hotly for the best deals, not only were start-ups getting more money right off the bat, but they were giving up less of their equity to do so. Gone were the days when Ken Olsen of DEC gave up a whopping 70 percent of his company for a mere $70,000. When Cambridge biotechnology start-up Repligen, whose scientific advisory board included two Nobel laureates, completed its second round of financing in 1983, it had raised more than $10 million and given away only about 20 percent of the company.

Similarly, when showy Encore Computer raised $22.5 million in a 1984 private placement, it sold only about 22 percent of the company. "We understood the mechanics of raising money, and we had personal track records that encouraged the outside world to invest in us," declared Ken Fisher. "We didn't want to give up ownership to people who were not going to participate in running the company."[30]

With its emphasis on high-profile founders, big bucks, and a lot of press attention, the Encore deal would have seemed more at home in Silicon Valley than in the more subdued Boston area. Both venture capitalists and entrepreneurs based in Boston agree that Route 128 start-ups tend to be less flamboyant and less well financed than their West Coast counterparts. "The East Coast entrepreneur tends to be a niche market entrepreneur," explains venture capitalist Bill Egan, a partner at Burr, Egan, Deleage & Company. "It's the Viet Cong approach to investing. 'We'll grab this little hamlet and we'll do a good job.' The West Coast mentality is, 'Listen, we'll get eight guys together, we'll announce in *Electronic News* that we're leaving and that we're overfunded, but you can give us a call if you want to give us more money.' Then they go after IBM or AT&T."[31]

This period of intense excitement and speculation in the venture community caught the attention of corporate America. According to Venture Economics, industrial companies nationwide increased the number of formal, in-house venture capital programs they ran from 28 in 1981 to 76 in 1986. Several prominent high-tech companies in the Boston area joined this trend. The goals of these investment programs varied markedly from company to company. Raytheon, Lotus, and semiconductor-maker Analog Devices, for example, viewed themselves as "strategic investors," while EG&G, which created its first venture fund in 1979, operated in the classic venture capital mode.

## GOVERNMENT'S ROLE

The federal increase in defense contracts during the Reagan years was clearly an important factor in the state's strengthening economy. There were other ways, however, in which government—particularly at the national level—was an instrumental player in Massachusetts's recovery. In 1977, for example, the National Science Foundation introduced a novel program designed to encourage the further development and application of the basic research it was funding. Although the agency had previously directed all its support to nonprofit institutions, it decided to require that a portion of its research grants—amounting to less than 1 percent of its total budget—be awarded to small business. Under the Small Business Innovation Research (SBIR) program, small businesses could submit proposals requesting funds to conduct advanced, high-risk research in areas showing commercial promise. Businesses that continued to meet the program's benchmarks could receive funding in up to three phases of development over several years: technical feasibility, follow-up research, and development for application.

The NSF was so pleased with the results that it joined with the Small Business Administration in pushing for a similar policy in other federal agencies with significant research budgets. The response was almost unanimously negative. "There was enormous opposition from every quarter you could imagine," recalls Roland Tibbetts, the former entrepreneur and venture capitalist who had devised the original program. "The defense industry, the OMB, everyone testified against it."[32]

According to Tibbetts, much of the opposition stemmed from doubts about the effectiveness of parceling out research to small firms without established reputations. Major contractors, of course, feared losing business to smaller competitors. Despite the outcry, Congress passed the Small Business Investment Development Act in 1982 by an overwhelming majority in both houses, largely because only a small amount of money was involved, and because congressional representatives were under pressure to support the small business sector.

The Act required eleven agencies to establish SBIR programs, including the NSF, NASA, the Environmental Protection Agency, the Nuclear Regulatory Agency, and the Departments of Defense, Agriculture, Commerce, Education, Energy, Transportation, and Health and Human Services. Each organization was required to earmark 1.25 percent of its external research budget for small business. Firms vying for awards had to propose a way to satisfy a specific research need selected from the lists prepared annually by each agency.

Ironically, the Department of Defense, which had been one of the fiercest opponents of the plan, has become one of its most ardent supporters. Massachusetts, with its finely tuned academic–industrial–military research complex already in place, has been a key beneficiary, ranking only behind California in the total number of contracts awarded. In addition, the program has encouraged businesses to collaborate with universities, and over half the awards involve such links. Since the inception of the program at the NSF, MIT has participated in a greater number of such partnerships each year than any other university.

According to local entrepreneurs, SBIR has been a boon to the area's research businesses, helping them to find a ready market for their expertise, while reducing some of the risk of conducting leading-edge research. "We used to spend a lot of time doing unsolicited marketing and trying to anticipate R&D needs," says Robert Weiss of Physical Sciences, Incorporated, who claims that at least 25 percent of his business is SBIR work. "Now we don't do that because we have a way to get our ideas in their earliest stages to the attention of someone with money, and with little investment on our own part."[33]

At the state level, it is clear that during the critical period of the 1950s through most of the 1970s, Massachusetts's government did not play a major role in the development or maturation of Boston's high-tech community. The network of small, research-intensive companies that arose built on the strengths and ideas flowing from the academic complex; on research funding from the federal government; on start-up money from the emerging venture capital industry; and on the companies that had come before them. Massachusetts did nothing overtly to encourage—or discourage—the high-tech community during those early years. Not until 1976 did the state draft a comprehensive economic plan.

As part of this effort, Dukakis created the Capital Formation Task Force, charged with developing new initiatives to help meet the capital needs of young businesses in the state. Among the outgrowths of this was the MTDC, a public venture capital corporation created in 1978 whose ambitious goals—above and beyond providing money to companies that could not find it in the private sector—included creating high-tech jobs in Massachusetts, attracting private investment, and nurturing entrepreneurship. Initial funding of $2 million came from the federal Economic Development Administration; the group made its first investment, in solar energy company Spire Corporation, in 1979.

Even as the state was launching such high-tech start-up-oriented initiatives, however, the high-tech community was banding together to combat both high taxes and what it perceived as an antibusiness climate in the state. In October 1977, irate executives formed a lobbying group dubbed the Massachusetts High Technology Council (MHTC). Membership was limited to the highest officer of participating companies or business units, with the exception of DEC, whose founder, Ken Olsen, did not care for politicking. In its first year, the council's ranks doubled in size, as the group hastened to make its presence known. "Before the council got together, there was no interchange between the companies here," recalls Ray Stata, founder of Analog Devices. "The presidents didn't know each other and there was no communication. Belonging to the MHTC has led to a networking among CEOs and a communication that never existed before."[34]

Although the MHTC did not take a stand on the gubernatorial race of 1978, many members made the defeat of incumbent governor Dukakis a top priority. The antagonism that existed between the governor and the business community was no secret. Dukakis's alleged insensitivity to corporate issues and concerns, coupled with his ardent support for far-reaching measures to boost social services and to protect the environment, had angered many chief executives. Helped by his "probusiness" platform, and, no doubt, by the financial support of many MHTC members, Democratic challenger

Edward King unseated Dukakis in a startling upset primary victory, and went on to win the election.

Under the King administration, the high-tech community appeared to make some inroads in changing the state's anti-business reputation, at least on paper. King won points with the young group by appointing MHTC member George Kariotis, chairman of Alpha Industries, as his Secretary of Economic Affairs, and by naming Stata and two other MHTC members to the Board of Regents for Higher Education. In addition, the governor in 1979 signed a nonbinding agreement drafted by the MHTC in which member companies promised to create 60,000 new high-technology jobs, and an additional 90,000 jobs in manufacturing and support services, if the state would cut taxes and work to improve the business climate. Although this sounded impressive, in fact, the lobbying group's members had already planned to add that many jobs before the "contract" was signed.

The MHTC really established itself as a force to be reckoned with, however, when it helped win passage of the so-called Proposition 2½, the 1980 tax-relief referendum that set a cap on property taxes at 2.5 percent of fair market value within each city and town. The MHTC believed that the measure, drafted by Citizens for Limited Taxation, a grass-roots organization of taxpayers, would lower taxes to the point where it would be easier to attract and hold employees.

Thus, by the late 1970s, members of the high-tech community had begun to form trade associations to discuss common interests and to boost their political clout. Bankers, lawyers, public relations firms, and real estate companies were tailoring their services to meet the special demands of these increasingly powerful and prosperous clients. In addition, the state initiated a series of programs to recognize and encourage technological growth, and to spread the benefits of high tech more equitably across the state.

The upshot of all this? Critical mass: an immense and complex network of interconnected and interdependent goods and services that had become almost self-sustaining. "The universities are not the magnet now," declared Bill Poduska, founder of Stellar Computer. "The mecca syndrome is what has happened. There are a lot of people here because there are a lot of companies here."[35] Added Encore's Ken Fisher: "The combination of venture capital, great schools, and a heritage of successful companies now feeds on itself."[36]

# 5
# The Calm Before the Storm

DRIVING ALONG ROUTE 128 TODAY, or walking through the Kendall Square area of Cambridge at the border of MIT, it is hard to imagine a Boston without high technology. The high-tech community seems omnipresent, infiltrating almost every aspect of life in the Boston area.

The creation of this incredible complex took decades. The universities generated the ideas for new businesses; the federal government concentrated personnel and money on specific projects; and investors put their faith and dollars behind hundreds of risky start-ups. At the heart of this network were the people who created it. The story of how all the strands of the high-tech network wove together into a coherent pattern of innovation began with the entrepreneurs who had the guts, the vision, and the perserverance to try an idea out on the marketplace and, sometimes, to make it work.

## THE FOUNDERS

Many people consider Ken Olsen to be the quintessential New England entrepreneur, and the very personification of the Yankee Puritan ethic. This is the man who said, "Success and wealth deteriorate your entrepreneurial spirit and your need to work hard," and, "You learn nothing unless it's painful."[1] The Boston high-tech community has encompassed a broad range of personalities, however, from the intensely private Olsen to the more gregarious and politically active Bill Poduska; from veteran hacker Richard Greenblatt, the former model-railroad enthusiast and founder of failed Lisp Machines Incorporated, to astute marketer and strategist William Foster of Stratus Computer.

A great number of local entrepreneurs first came to the Boston area to attend MIT. A 1989 Institute study of ninety-nine graduates who started companies in Massachusetts presents an interesting profile: More than half of these entrepreneurs had parents who were themselves entrepreneurs and almost 75 percent had earned advanced degrees. Their families had an average of three children and their most favored hobby was tennis. The most frequently cited reason for starting a company was to be independent, followed by the desire to become financially successful.[2]

Many other entrepreneurs came to the area, drawn by opportunities at Raytheon, DEC, Wang, and other top employers. Once here, the specific drive or motivation that prompted each to launch a company was uniquely his or her own.

Amar Bose, for example—tall, silver-haired, and charismatic—set up a home radio repair shop in Pennsylvania as a thirteen-year-old to help support his family when his father's small Indian import business had to shut down during World War II. At the shop's peak, Bose employed seven fellow students, paying them fifty cents an hour. "Everybody who could repair radios was in the service at that time," Bose explains. "This grew to be the largest repair shop in the Philadelphia area during the war. I closed it only in 1947 to come to MIT."[3] After graduation, Bose became a professor at MIT, and began his experiments on loudspeaker design that led to the founding of Bose Corporation, the Framingham-based maker of state-of-the-art audio products.

George Hatsopoulos showed similar industry at an early age. "I always felt that I wanted to do entrepreneurial things and go into business in technology," he recalls.[4] At the age of fourteen, living in Greece, Hatsopoulos displayed entrepreneurial spirit and boldness by ignoring a ban on radios imposed by the Germans during their 1941 occupation. The young boy began building radios in his family's garage and selling them, against the German decree, mostly to the Underground. "My father found out about my activities in 1944, a few months before the Liberation," Hatsopoulos chuckles. "He got so upset that he made me take all my tools and all the components I had and then we went up in the mountains and dug a hole and buried them."

Hatsopoulos's brainstorming activities proved to be more long-lived than his tools. "Throughout the World War, I was thinking of inventions and I had a notebook where I kept all these ideas," he recounts. "My intention was when I finally finished school, I would use one of these inventions to start a business." When Hatsopoulos prepared to start his doctoral thesis at MIT in 1954 he went back to his notebook—which he still keeps at home—and looked for an idea that would provide both a good topic and the basis for a company. There he found a page describing thermionic emission dating back to 1942 on which he had scrawled, "Could be used one day to convert heat into electricity without moving parts." Hatsopoulos methodically went to work to achieve his plan. In 1956 he received his Ph.D. in mechanical engineering, was appointed an assistant professor at MIT, and formed Thermo Electron, all in the same month.

Others who took up the challenge of building a company did so less because of an entrepreneurial drive than to achieve a specific goal: doing the work that most interested them. Alexander d'Arbeloff, for example, started Teradyne partly out of self-preservation after being fired from three jobs in four years. "I had a lot of ideas, and I just didn't have a lot of patience," the son of Russian immigrants notes wryly. "It's interesting, by the way, that the three companies that fired me no longer exist."[5]

Kenneth Fisher was ensconced as vice president of central operations at Honeywell Information Systems in Chicago in 1975 when his chance to head a business arose. "I thought of myself as a big company guy, running a major part of a big company, and working my way up the ladder as far as talent and circumstance allowed," he muses. Then a headhunter called and asked Fisher if he'd like to run a computer company. "I thought, 'OK! Finally I've been recognized!' I'm thinking Burroughs or NCR. I said, 'Yeah, I'll talk. Who is it?' The guy said, 'Prime Computer.' I said, 'Who?' I'd never heard of them. And ninety days later I was the president."[6] In six years, before resigning over disagreements with the company's board of directors, Fisher transformed Prime from an engineering-driven company with 1975 sales of $11.4 million to a savvy marketer with 1981 sales of almost $365 million.

Thomas H. Fraser, a founder of Cambridge-based Repligen

Corporation, like Fisher expected to remain a loyal employee of an established corporation. From 1977 to 1981, Fraser directed the genetic engineering research program at the Upjohn Company in Kalamazoo, Michigan. "My goal was not to become a millionaire or to have the power of running my own company or anything else," he insists. "My goal was to create a world-class, cutting-edge technology unit."[7] Fraser made the decision to return to Cambridge, where he had earned his Ph.D. from MIT in biochemistry, after growing disenchanted with Upjohn's pace of expansion. "Unfortunately, a large company like Upjohn's idea of growth and mine are not the same," he grins. "I hired between one and two people a year there." Fraser founded Repligen in 1981 and left in 1990.

Mitchell Kapor, who started Lotus Development Corporation in 1982, says his entrepreneurial tendencies lay beneath the surface. "Several times in my early twenties I had little businesses," he recalls. "In college there used to be these bootleg Bob Dylan albums, and I was the local campus distributor for awhile. Then I started a company with a partner producing new radio features. But I never thought about going to the Harvard Business School while I was growing up."[8] Early work experiences included stints as a disc jockey, a Transcendental Meditation instructor, and a health care worker.

Kapor later attended MIT's Sloan School of Management, but dropped out before completing his MBA. He eventually decided to take the big step of founding Lotus, against the advice of many of his friends, because he couldn't find enough publishing opportunities as an independent software creator. "Some part of me that I could not acknowledge consciously at the time really wanted to start my own business, and the reason I could not acknowledge it is I was too scared that if I failed,

I couldn't live with myself," he explains. "So I had to go through this rather circuitous pretense of talking myself into it." Kapor left Lotus in 1986 a multimillionaire, and about a year and a half later launched a second software start-up, ON Technology Incorporated.

The oft-reported model of how to start a high-tech company usually follows a reasoned and methodical plan. The entrepreneurs come up with a specific idea or innovation, prepare at least a rudimentary business plan, look into the availability of some initial venture capital, and only then quit their jobs to launch their new ventures. The reality, however, is often quite different.

Bill Foster, a California native, worked his way up the corporate ladder at Hewlett-Packard and Data General through the 1970s. By 1979, he was vice president of software engineering at DG. He was also convinced that he wanted out. "I wanted to become president of a computer company—that's what it boiled down to," he recalls. "I got to know the presidents of a lot of computer companies, both big and small, and I slowly figured out that these people didn't walk on water. The only difference between them and everybody else is they were willing to accept the possibility of failure."[9]

On a business trip to Europe for Data General in June 1979, Foster woke up abruptly one night and decided to quit. Within three weeks of his return from overseas, he had resigned. "I had no idea about how to raise money, or even about what was available," he now admits. "I had no financing, no partners, no ideas when I quit. Most people thought I was crazy. I probably was." Despite this dubious beginning, made even more inauspicious by a failed attempt to locate the new company on the start-up-weary West Coast, Foster's Stratus Computer Incorporated, founded in 1980, flourished.

It has been lauded as a particularly well managed and keenly focused competitor.

Like Foster, Paris-born Phil Villers, a man known for his support of liberal causes, decided to start a company before he knew what it would be. A graduate of both Harvard and MIT, he was thirty-four years old when he embarked on CAD/CAM-maker Computervision. "The immediate timing of the founding of Computervision, oddly enough, was connected with the death of Martin Luther King," he says. "There was a very moving sermon on his life preached in our church on the occasion of his death and one of the things that struck me was how much he had accomplished at an age just slightly older than I was then. I realized that if I was going to do something about my lifelong ambitions, now was the time to get moving, and, in fact, I started planning Computervision that very afternoon."[10]

"Now or never" was a common refrain among entrepreneurs. Ronald Gruner gained a certain notoriety through the best-selling *Soul of a New Machine,* which included the account of a failed bid by a North Carolina team, of which he was the leader, to create Data General's next-generation minicomputer. Gruner reached his turning point after fourteen years with DG. "When I joined Data General I was twenty-one years old," he recounts. "I told Ed de Castro at the time I joined that I expected to leave Data General when I was twenty-five and start my own firm."[11]

Gruner's estimate was off by ten years. At the age of thirty-five, while recovering in the hospital from an emergency appendectomy, he decided that if he were ever going to start a company, this was the time. Gruner and co-founder Craig Mundie, also from Data General, incorporated and named their company (later renamed Alliant Computer Systems Corporation) before they had even settled on the idea of making parallel processing computers. Gruner left Alliant, which was experiencing hard times, in 1991.

Many of these entrepreneurs had ties to MIT, but most would not fit the stereotype that has evolved of "hackers," the nerdy computer nuts who purportedly cluster at places like MIT, and who are so absorbed in their work that they lose touch with much of society. One man who does fit the role of hacker-turned-entrepreneur, though, is Richard D. Greenblatt, who had the honor of being featured in Steven Levy's book, *Hackers: Heros of the Computer Revolution.*[12]

Greenblatt came to MIT as a freshman in 1962, but he became so absorbed with computers and the Tech Model Railroad Club—whose members pioneered fundamental computing concepts with their sophisticated train set—that he flunked out his sophomore year. Greenblatt's association with MIT was just beginning, however. The plump and disheveled computer whiz returned to the Institute's Artificial Intelligence Lab in 1965 and remained there for more than fifteen years, numbering among his accomplishments the creation of a canny computer chess program and groundbreaking programming work in LISP, now the computer language of choice in artificial intelligence applications.

Although widely acknowledged as a brilliant programmer, Greenblatt did not see eye-to-eye with most of his associates on business matters. In 1979, when the LISP machines that the AI lab was producing for its own research began attracting attention both inside and outside the Institute, the group behind the project met to discuss forming a company to market the computers. According to Russell Noftsker, who had worked as general manager of the AI Lab for eight years before starting a company in California, and who had returned to drum up interest in commercializing the LISP machine, Greenblatt's idea of a company and the rest of the group's diverged wildly. In essence, Noftsker maintains, Greenblatt's concept remained true to the vision of a hacker.

"His view at the time was that there should be no outside employees, there should be no payroll, that they would build the machines at night and on weekends, and that no one would get any salary," Noftsker contends. "Of course, there was no consideration of overhead items, marketing, or sales. And Richard would be in charge of everything. We held that organizational meeting in February 1979 and tried to come to an agreement, because all of us felt that Richard had played such an important role in this and deserved to be part of it. But we also all knew that we were not going to run the company the way he wanted to run one. We had a very long and traumatic meeting at which we discussed all these things, and Richard maintained his insistence that he be in charge of everything because it was his baby, his product. We weren't able to convey the impression to him that we couldn't accept that."[13]

After a vote of no confidence in Greenblatt's plan, Noftsker says, the rest of the group agreed to give Greenblatt one year to start a company. If he hadn't done it by then, they would launch their own AI start-up. A little more than a year later, a group of twenty-one founders headed by Noftsker founded Symbolics Incorporated, a pioneer in artificial intelligence technology. A few months after that, Greenblatt made his own entrance into the business world with the competing Lisp Machines Incorporated.

Greenblatt declines to discuss the details of that original breakup. "There were some personality issues and there were some questions of approach," he mumbles, dismissing the subject. "There were different things that divided us."[14] Seven years later, Greenblatt and Noftsker had more in common: companies beset by internal problems, fierce external competition, and a market that had not materialized so quickly as most AI enthusiasts had predicted. In 1987, Lisp Machines went bankrupt. As for Symbolics, after sales reached $114 million in 1986, the company's fortunes went downhill. The board of directors ousted founder Noftsker in January 1988. Symbolics relocated to the outlying town of Burlington, and the company's business focus shifted from hardware to software.

As these examples illustrate, most start-ups—even those with brilliant and promising starts—didn't achieve the success or stature of a Stratus, a Lotus, or a Thermo Electron. This sobering reality didn't deter most entrepreneurs, however. Moreover, although many of those who started companies undoubtedly dreamed of being the next DEC or Biogen, others insist that they are more easily satisfied. "We haven't experienced mega growth, but for what we're interested in, we've carved out a niche where we're the clear leader," declares Peter L. Gabel, who left Lotus Development in 1984 to found Arity Corporation, a small maker of artificial intelligence software. "We still are going to be a very successful company, but our time frame is different."[15]

## SPINNING OFF

Arity, in being a spin-off, is typical of hundreds of resourceful companies in the Boston area. The state's high-tech community was largely built on innovative ideas—sparked by the interactions of people and organizations—being spun off into new products and enterprises. As we have already seen, the joint efforts of the federal government and MIT during World War II spawned hundreds of new products and companies.

Moreover, companies spinning off from other companies were at the very heart of the monumental growth that the Route 128 area experienced from the 1960s through the 1980s. Often, no sooner did a company establish the viability of the marketplace it was pursuing than key employees began breaking loose to try and do it better on their own. In most cases, these fledgling entrepreneurs stayed in the neighborhood, relying on the same network of customers, suppliers, and friends who had supported them in their previous jobs.

Tracerlab was one of the area's earlier spin-off incubators. This small entrepreneurial firm, founded shortly after the conclusion of World War II by a group formerly associated with MIT's Radiation Laboratory, was among the earliest companies to develop peacetime uses for nuclear science, including reprocessed radiochemicals and nuclear instrumentation. Both because it was one of the first companies backed by the fledgling American Research and Development venture capital company, and because its technology was seen as unusually promising and glamorous, Tracerlab garnered extensive media attention. A lengthy profile in the December 1947 issue of *Fortune* magazine glowingly described Tracerlab as, "The first real business to be built entirely out of byproducts of the atom bomb. . . . It was born out of the atom bomb by radar, and thus may claim direct descent from the two biggest scientific projects in all history."[16]

A number of factors, however, including a divisive relationship between top management and the company's technical staff, a misguided acquisition that left the company short of funds for other development efforts, and low salaries, caused Tracerlab employees to defect at an alarming rate. Within six months of the company's incorporation in early 1946, four of the original six founders had moved on to other jobs. In the sixteen years before it was acquired and disassembled, Tracerlab was the source of nineteen spin-offs, thirteen of which used technology taken directly from the parent company. One of the most successful, New England Nuclear, had sales of close to $100 million in 1980 when Du Pont bought it for $342 million.

Tracerlab was not alone. In a 1981 thesis, MIT master's candidate Christopher L. Taylor set out to identify and study the spin-offs from EG&G, whose broad range of distinct technical businesses made it a likely candidate for spawning new firms.[17] Taylor's initial list of fifteen spin-offs soon snowballed to fifty-eight companies in New England, and an estimated total of more than seventy-five nationwide. In his study, Taylor found that, on average, the entrepreneurs had contemplated striking out on their own for nine years before actually doing it, and that the cause that most frequently precipitated the move was a change in work assignment, followed by frustration on the job. As in the study of MIT graduates who started their own companies, the features most commonly cited for striking out on their own were the desire to be their own boss, and the wish for financial reward.

One of the region's better known spin-offs is Data General, whose founder, Edson de Castro, made an acrimonious break from Digital Equipment with two other DEC engineers in 1968. The soft-spoken, then-thirty-year-old de Castro—who had a reputation within DEC as both a bright engineer and a rebel intent on doing things his own way—left DEC in a swirl of controversy after Ken Olsen rejected his design for an ambitious new minicomputer. Although Olsen never brought suit against the upstart Data General, he was reportedly infuriated by what he saw as the group's disloyalty and its dishonesty in designing a prototype for their new company's computer while still employed at DEC. Eleven years after the defection, Olsen told *Fortune* magazine, "What they did was so bad we're still upset about it."[18] De Castro, for his part, insists that Data General scrapped those original plans and designed a better machine that was soon selling head-to-head against DEC.

"We felt that the industry was really just in its infancy and that the opportunities were really enormous," de Castro explains of his departure. "It was a question of opportunity and wanting to get into something that we felt was going to explode."[19] In 1991, more than twenty years after DG's founding, the two founders were still cool toward one another.

Ironically, for a company whose own spin-off and creation created such a stir, Data General gained quite a reputation as a parent of spin-off firms itself. The *Boston Globe* identified at least seventeen companies founded by Data General veterans during one particularly tempestuous four-year period in the early 1980s, and there were others before and since.[20] With Foster's Stratus Computer and Gruner's Alliant Computer, such heavyweights as Apollo Computer, purchased by Hewlett-Packard in 1989, and software maker Lotus Development, have been co-founded by ex-DG employees.

In addition to Data General, start-ups founded by ex-DEC employees include now defunct computer makers Masscomp and Mosaic Technologies. Prime veterans have helped launch such ventures as Apollo Computer and Raster Technologies. Apollo itself has spawned at least nine companies. Encore Computer's founding team was comprised of top talent from all three of the leading minicomputer makers—Ken Fisher from Prime, Henry Burkhardt from Data General, and Gordon Bell from DEC. Lotus Development, within three years of its founding, had spawned at least three software companies, Iris Associates, Arity Corporation, and Jonathan Sachs Associates. Charles Francis Adams, former chairman of the board of Raytheon, still grumbles about the mutiny of eleven engineers who broke off to form defense electronics competitor Sanders Associates Incorporated—now based in New Hampshire—in 1951.

The relationship between a parent company and its spin-offs has ranged from out-and-out war to financial support, depending on how critical a role the spin-offs' founders played at the parent company, and on what special knowledge and technologies they took with them. Tracerlab's Bill Barbour, for example, is still bitter some forty years after the fact about his failure to stop, or to effectively retaliate against, former employees who set up businesses in direct competition with Tracerlab. On the advice of Robert Luick, Tracerlab's legal counsel, Barbour never took legal action against any of the spin-offs, though, in retrospect, he believes he would have been successful.

Whether an employee leaves his parent company with a blessing or a curse, the first years of managing a start-up are never easy. Being an entrepreneur almost invariably requires a dedication that goes far beyond a sixty-, eighty-, or even one hundred–hour week. Raising money, hiring competent and compatible employees, and getting a product out the door on schedule with limited help and resources can stretch even the most hard-charging executive to the limit. For some entrepreneurs, apparently, it can also be an exhilarating roller coaster ride that is hard to get off.

These are the repeat entrepreneurs, people like Bill Poduska, Phil Villers, and Raymond Kurzweil. Often these are also people who recognize, or are forced to admit, that their own strengths lie not in the day-to-day operation of an established company, but in the more creative and visionary process of getting an idea off the ground. Although Poduska, Villers, and Kurzweil are among the area's most visible repeat entrepreneurs, starting more than one company is surprisingly common. According to the MIT study of ninety-nine founder-graduates, the average number of companies founded by these entrepreneurs was 1.9, or almost two companies apiece.[21]

Next to getting financing, hiring quality employees who can carry a company forward and fill the ranks as a company grows is one of the hardest tasks facing an entrepreneur. Many have handpicked a solid team at the start but that does not solve the problem of where to find the engineering, managerial, production, and clerical workers an expanding company requires to grow. The Boston area throughout the 1980s had one of the lowest unemployment rates in the entire country and companies trying to hire there faced the additional obstacle of an overall shortage of workers. In 1989, for example, fast-food restaurants around Boston were so desperate for help that they offered high school students starting wages of up to $6.50 an hour.

Regardless of the quantity of potential employees, the quality remained intact. Even today, the depth of talent available in the Boston-area work force is unmatched anywhere in the country, except, perhaps, in California. Bill Foster of Stratus likens hiring in the area to a food chain. "There's kind of a cycle you see here," he explains. "The most critical thing in starting a computer company is being in an area where there are a lot of big computer companies so you can draw experienced people away from them. And the big computer companies need to locate in an area where there are a lot of schools so that as they lose people to the start-ups, they can replace them with people fresh out of school." He adds: "I see that as the key to the whole food chain for the Route 128 area. If the big companies weren't here, we wouldn't be here, and if the schools weren't here, the big companies wouldn't be here."[22]

Mitch Kapor, founder of Lotus and ON Technology, seconds Foster. "We're better off here in terms of software people than we would be on the West Coast," he declares. "There are some other areas that are up-and-coming, like Boulder, Denver, and Austin, but in 1981 when Lotus really got started, it was either here or California."[23]

Although the sixty-five colleges and universities in the greater Boston area—and in particular such science- and engineering-oriented schools as MIT and Northeastern—have been prime suppliers of talent to the state's high-tech industry, not all companies can use engineers or scientists right out of school. Some smaller companies, in particular, hire just a few highly specialized and experienced professionals. During the 1980s, however, many high-tech firms relied on the ability and drive of recent graduates. Bill Bowman, founder and former chairman of Spinnaker Software, says the company's Cambridge location provided an excellent labor pool. "We decided to hire smart people regardless of their experience, and there are a lot of smart people around here," he says.[24] Similarly, Phil Villers drew on MIT graduates for all his start-ups. "All three of my companies have been pioneers in an important new area of computerized systems, so for a very bright young graduate, this was a plum assignment," he explains. "You need a combination of brilliant new graduates and seasoned engineers."[25]

## SERVICE AND SUPPORT

As the high-tech community began to flex its political muscle during the 1970s and early 1980s, it also began to reap the benefits of its considerable and growing financial power. No longer just a random scattering of aggressive, high-growth companies, the Route 128 crowd was now big enough, diverse enough, and successful enough to attract a multilayered infrastructure of companies and organizations anxious to win its business.

Part of this infrastructure arose almost from the start. As soon as a Raytheon or a DEC developed production needs large enough to reach outside the company, small companies adapted or sprang up precisely to fill those needs. By the time Stratus Computer started up in 1980, there was a well-established network of suppliers serving the Boston-area electronics community—another critical component in founder Bill Foster's decision to locate near Route 128. "The thing that's helped us a lot are not so much the peer companies in the area whom we tend to compete with, but the service companies that are around," explains Foster. "People that do our sheet metal for us, people that produce our PC boards, or who stuff components on the PC boards, that's all done by local companies." He adds: "Having them local is very important, because if you have a quality problem, you want to be able to drive over quickly and see what's going on."[26]

The kind of support that start-ups in the area came to take for granted went far beyond machine shops and component suppliers. Virtually every type of service organization in the area—including banks and law firms—made it a top priority to anticipate the needs of the high-tech sector, and to tailor their products to these typically small, high-growth companies. These services were particularly important for the start-up that did not have the resources to maintain its own in-house legal or accounting staff. "You can find good law firms, accounting, and people who understand physical facilities for high-tech people," declares Mitchell Kapor of Lotus and ON Technology. "That is a major advantage when you're trying to grow very, very rapidly."[27] Adds Ken Fisher: "I grew up in Omaha, Nebraska, and I go out there and think, I could never have started Encore in a place like this. I couldn't even find a lawyer who would know how to incorporate the company properly."[28]

Incorporating companies was just the start for attorneys specializing in Route 128 companies. Although the moniker "high-tech lawyer" might seem contrived, these lawyers offered skills that a traditional industrial company would not require. A high-tech lawyer might help advise an entrepreneur on everything from putting together a business proposal to copyrighting a software product to arranging an R&D partnership.

The major accounting firms also courted high-tech customers. Price Waterhouse, for example, opened an office in 1986 in Cambridge's Kendall Square to tap that area's rich concentration of growing, technology-based firms. This Entrepreneurial Services Center, which is geared toward assisting small, emerging companies, augments its marketing materials with software discs on such topics as "Developing the Business Plan for Your High-Technology Company" and "Tax Planning Opportunities for High-Technology Companies."

Miller Communications Incorporated, a Boston-based public relations firm founded in 1977, switched to handling only high-tech accounts early in the 1980s after the phenomenal success experienced by two of its clients, Lotus Development and Houston-based Compaq Computer Corporation. "All of a sudden we were hooked to the tail of two comets, and the rest is history," declared partner Frank W. Morgan. "The high-tech companies lined up outside our door. We went from $300,000 in billings in 1982 to $5 million in billings in 1986."[29]

The emergence of this extensive support network was gradual, tied directly to the late-blooming realization of the financial impact high tech had on the region. Aaron Kleiner, vice chairman of Kurzweil Applied Intelligence, Incorporated, remembered when things weren't so easy. "What's happened in the last ten years is that the whole community supports start-ups and understands start-ups as an opportunity rather than as something to be avoided," he explained. "There's been a total shift not just in the availability of money but in the availability of real community support for the process. Before, if you didn't work for Raytheon, you were nothing."[30]

## NETWORKS AT WORK

During the 1980s, the different segments of the Boston area's high-tech network—the entrepreneurs, the universities, the skilled workers, the lobbying groups, the service organizations, the state programs, and the high-tech companies themselves—were like pieces of a complicated puzzle. Each segment was necessary to make the picture complete. Together, they formed the particular environment of the Route 128 area, an environment that both nurtured the creation of new high-tech start-ups, and drew existing companies from across the country, and even from overseas. By 1965, out-of-state companies with outposts along Route 128 already included Sylvania Electric Products, Honeywell, and Hewlett-Packard. Over the years, as the region's reputation grew, so did the influx of outsiders.

One of Japan's computer giants, NEC Corporation, headquartered a major subsidiary, NEC Information Systems Incorporated, in Lexington in 1977, and later moved it to Boxborough. In 1988, Bull HN Information Systems Incorporated, the computer company owned by Groupe Bull of France, NEC of Japan, and Honeywell, relocated its U.S. headquarters from Minneapolis to Billerica, northwest of Boston. And in 1991, Japanese giant Mitsubishi Electric Corporation announced it would build a 100-person basic research laboratory in Cambridge's Kendall Square.

That the Route 128 high-tech network has made its mark seems undisputed. The high-tech community has achieved a critical mass capable of sustaining a strong economy without outside interference. This intricate, sprawling network, and not any particular geographic boundary, best describes the Route 128 community. Despite the spontaneity of its creation, the vigor of its growth, and the nurturing effect of its powerful connections, however, this network has existed in a delicate balance. Moreover, its very complexity has masked the effects that changes in economic or political forces could have on the process.

## DISCONNECT

Although the "Miracle Years" in Massachusetts seemed a time of widespread prosperity and promise, not every sector did equally well. The state continued to struggle with such chronic problems as homelessness in its urban centers, a lack of jobs for inner-city youth, and an inability to bring all the Commonwealth along for the ride. Nevertheless, the state did considerably better than the nation as a whole in addressing these issues. Liberal social service policies and commitment to "spreading the wealth" succeeded in closing some of the traditional gaps that had existed between rich and poor, and white people and those of color. Middle-income families saw their standard of living rise, jobless rates for minorities fell sharply, and state-sponsored business initiatives at aging industrial centers such as Fall River and Lawrence created new jobs and new hopes for the future.

As miraculous as this economic surge may have been, it was by no means invulnerable. The industry-driven boom eventually fell victim to a combination of changing technologies, aggressive global competition, economic setbacks, and greed. As early as 1984, there were signs that the venture-financed high-tech explosion had gone off course, and by the late 1980s, venture capital firms had begun to shift away from the classic meat and potatoes of venture investing—the small, high-tech start-up. Venture firms put an increasing percentage of their funds into different kinds of deals, ranging from leveraged buyouts to investments in nontechnical companies. In addition, many of the high-tech investments were made in more mature companies that had already exhibited their basic competitiveness and ability to survive. In 1987, although the amount of venture investments soared to $4.9 billion, the percentage spent on early stage investments dropped to 28 percent, down from 37 percent in 1984. Similarly, in 1990 only 27 percent of the almost $2 billion invested went to such early stage financing.

This shift away from start-ups left many industry observers predicting a fallout in innovation, with entrepreneurs struggling to find backers still interested in taking a chance. To a certain extent, this prediction was borne out. Entrepreneurs starting companies in the mid- to late 1980s typically did not find the same easy access to money that their counterparts of five years earlier had enjoyed. "It took us a year to raise $150,000 from seven private investors," complains David A. Blohm, who co-founded Cambridge software maker MathSoft Incorporated in 1985. "In 1980 we could have gotten that in a day and a half."[31]

But the sheer size of the funds that venture capitalists continued to raise, despite adverse conditions, made the late 1980s very different from the mid-1970s, when there simply was not much money around. In 1986, for example, a single new fund, E. M. Warburg, Pincus & Company, Incorporated, raised $1.2 billion, setting an industry record and boosting New York to first place in total venture capital funds committed that year. By 1990, the total pool of venture money was almost $36 billion.

As the 1980s progressed, the largely regional character of the venture business also changed markedly as such companies as TA Associates, Greylock, and Matrix Partners opened offices on the West Coast, and Hambrecht and Quist and Merrill, Pickard, Anderson & Eyre—two Silicon Valley investors—opened offices outside of Boston. Stephen Coit, who heads Merrill, Pickard's Waltham operation in the heart of the Route 128 region, contends that having representatives on both coasts was no longer a luxury, now that competition had forced venture capitalists to become more specialized. "As soon as you focus on vertical markets, you have to give up on the geographical boundaries," explains Coit. "And one force behind specialization was the oversupply of venture capital."[32]

Specialization became one of the key strategies of the venture capital community. The more aggressive investors no longer waited for entrepreneurs or other venture capitalists to bring them proposals. Coit, for example, "walks the walls" at trade shows looking for undiscovered talents in the fields of life sciences, communications, and semiconductors. To help track down people and ideas in the medical arena, Coit has had a molecular biologist working part-time as a consultant on exclusive retainer. Once a deal is underway, Coit often calls in an industrial psychologist and a headhunter to help recruit and evaluate potential management candidates. "It used to be you'd sit around in the club puffing on your pipe in an over-stuffed chair and say, 'Is there any way you can find some money to put into this company?' and the other guy would pull out his pipe and say, 'I believe I can spare $500,000,' " Coit says. "That's not the way it's done anymore. If those clubs still have guys sitting there, they are venture capitalists with shrinking portfolios."

The changed face of venture capital has had both good and bad implications for entrepreneurs. On the one hand, many venture capitalists have become so cautious and so specialized that they often will not even consider investing in a company unless it fits with their portfolio mix. Moreover, the trend for many investors at the end of the 1980s was to shy away from the early stage companies that require more patient nurturing.

On the other hand, the industry's very specialization could also work in favor of the small start-up. Several venture capital firms in the Boston area—including AEGIS Funds and Venture Capital Fund of New England—have chosen to concentrate specifically on the local entrepreneur looking for a relatively modest investment. "This area is unique in that it has such a broad diversity of technology and lots of little niches that can be attractive for us," explains Venture Capital

Fund's Dick Farrell.[33] Others, like Zero Stage Capital, cofounded by MIT professor Edward B. Roberts, specialize in seed capital and early stage investments in the $50,000 to $500,000 range. Moreover, although the end of the decade brought a more cautious approach to investing, some venture capitalists were still willing to go all out for a particularly promising technology.

Nonetheless, the start-up arena was one of the first places where cracks began to appear in the facade. For many people, the Massachusetts Miracle's prime importance was as a job machine: Scores of entrepreneurs were starting up new firms; the existing mainstay technology companies were hiring thousands of new employees to satisfy increased demand for their products; and the network of service and supply companies radiating out from the high-tech center accounted for the greatest job growth of all. As the 1980s unfolded, however, each of these sectors came under attack.

## THE BUBBLE BURSTS

To claim that greed alone was the beginning of the end would be overstating the case, but many people were apparently blinded by the hype and hoopla of the Miracle years. A "get-rich-quick" mentality spread among many would-be entrepreneurs, developers, and already well-to-do venture capitalists, inspired by the oft-quoted examples of people who had hopped on the high-tech engine and amassed personal fortunes in the process.

There was a dark side to the gold rush mentality of this period. It was so easy to raise money that many companies got started that in a normal investment climate would have been judged too weak to compete. Moreover, the lemminglike behavior of many of the venture capitalists, who rushed to include representatives of the latest hot technology in their portfolios, guaranteed that the market would be unable to support all the competitors struggling for market share.

Perhaps the most notorious example of this overkill, on a national scale, was the reckless investment in Winchester disk drives, high-speed data storage devices for computers. Venture capital firms invested almost $400 million in forty-three different manufacturers of Winchester disk drives between 1978 and 1984, and during the same period, disk drive manufacturers raised more than $800 million in public offerings.

Not surprisingly, the inflated valuations could not be maintained in such a competitive market. By 1984, the market values of the publicly held disk makers had plummeted, sales of many manufacturers were faltering, and some companies would soon drop out of the market altogether. "The old equation in the early seventies was if you picked a good entrepreneur in a growing market and got ten percent, you were in great shape," recalls venture capitalist William Egan. "What happened in the early eighties was there were ten companies all going after that ten percent of the business."[34]

Winchester disks were not the only high-tech stocks to lose favor. In 1984, the public invested less than $750 million in high-tech companies, down from more than $3 billion in 1983. Venture capitalists, too, became more wary. For the first time since 1979, total new capital committed fell in both 1984 and 1985. Not only was there no longer an eager public market, ready to snap up start-ups after just a few years, but many venture capitalists had their hands full dealing with "sick puppies"—companies that were not performing up to expectations, or were drowning in a sea of competition.

Personal software makers, which had crowded into the marketplace with such enthusiasm just a few years earlier, found the going particularly rough. In October 1984, Ovation Technologies Incorporated and Knoware Incorporated, two particularly promising and well-financed Boston-area start-ups, both announced plans to shut down within two weeks of each other. Knoware, whose first product was a program designed to teach professionals how to use personal computers through a learning game, was founded by two professors at MIT's Sloan School of Management and had received $2.5 million in venture capital when it filed for liquidation. Although Knoware's software received good reviews, co-founder Stuart E. Madnick claims that the management team spent more on marketing than the company's one product could support. "The company ate itself out of existence," Madnick concedes.[35]

Ovation Technologies never even brought a product to market. The Norwood company, which planned to go head-to-head with Lotus in the business software market, appears to have been, at least in part, another victim of the obsession with marketing and promotion. Ovation made a splash by running full-page ads in the *Wall Street Journal* and entertaining the press at the Windows on the World restaurant in New York. Yet the company was unable to resolve technical problems and get the software out the door, despite the fact that it had raised $6.8 million in private financing, an enormous sum for a software company.

Biotechnology companies also fell into disfavor among members of the investment community. Urged on by tales of the broad promise of biotechnology for curing diseases, boosting agricultural yields, producing chemicals, and providing new sources of energy, investors poured millions of dollars into gene-splicing start-ups in the early 1980s. Most of these companies turned out to be a longer-term investment than investors had anticipated, however. Not only did it take time and money to turn the basic science into real products and industrial processes, but meeting costly and time-consuming regulatory requirements in the areas of pharmaceuticals and agriculture delayed the arrival of products in those potentially lucrative markets. One investor finding itself in the hot seat was Boston University, which in 1987 bought a majority stake in Seragen, Incorporated. By the end of 1991, the university had sunk more than $80 million in the small Boston-area biotechnology firm, and critics both inside and outside the university decried the unusually large investment in a start-up that was by its very nature high risk.

On top of the problems of weak companies, congested marketplaces, and misunderstood technologies, Congress—as many investors had feared—eliminated the capital gains differential when it passed the Tax Reform Act of 1986, making capital gains taxable at the same rate as regular income. The change did not have nearly the impact of the hike in the capital gains tax back in 1969, primarily because the overall tax rate was lower, and capital from tax-exempt institutions such as pension funds and university endowments by this time made up more than 50 percent of the total investment pool.

Nevertheless, many critics claimed that the change sent a discouraging message to the thousands of budding entrepreneurs planning to take a risk on a new idea. "Anything that happens to dampen down that optimism will kill the process," declared William Egan of Burr, Egan, Deleage. "The fact that we don't have a differential between regular income and capital gains is a disgrace."[36] Moreover, the tax change hastened a shift in the kinds of investments venture companies were making. With less incentive to put their money in start-up investments—not only because of the capital gains change, but also because of the uncertain market for high-tech stocks, and the amount of money and time such investments require—venture capitalists in the last half of the decade began to seek out easier ways to earn high returns.

## HIGH TECH SINKS LOW

The Boston area had been the nation's first high-tech center and, with the exception of Silicon Valley, had been the country's unrivaled cradle of innovation. This historic prominence helped make the region's excellence self-perpetuating: The more famous the area's universities and companies became, the more bright and talented people were attracted to Boston and Cambridge, guaranteeing that the ideas, inspiration, and energy that lie behind innovation would continue to find a home there.

The fact that the Route 128–area companies were among the progenitors of high tech also had the potential to work against the region, however. Like Rust Belt companies, forced to retool and restructure to survive in the modern age, many of Massachusetts's high-tech companies suddenly found themselves in the unenviable position of feeling old—of lagging the technological leading edge. Nowhere was the displacement clearer than in the state's once booming minicomputer industry. Digital Equipment, Wang, Prime Computer, Data General, and Bull HN, key representatives of Massachusetts high-tech, now found their businesses on the line.

Although the state had been losing jobs from its high-tech manufacturing sector since 1984, serious problems did not begin showing up in the minicomputer industry until the end of the decade. Like other businesses, minicomputer makers were hit hard by the stock market crash of 1987, the weakening economy, and the nation's subsequent slide into recession. What hurt even more was a fundamental shift in technology: The minicomputer was out, and personal computers and workstations were in. In addition, most Massachusetts computer makers had embraced a flawed and outmoded strategy. Instead of assembling hardware using off-the-shelf parts, they had attempted to keep customers tied to in-house-developed machines and software that were not compatible with other manufacturers' products. As industry standards increasingly became the order of the day, these mainstays of the Massachusetts business community became highly vulnerable.

Looking back, it is startling to see how abruptly the Miracle faded. As Governor Dukakis's quest for the presidency—largely premised on the economic miracle in Massachusetts—began to derail in the summer of 1988, the state's fortunes also began to run off track. First came the news of an unexpected budget deficit, the first since 1975. Tax revenues failed to meet expectations, in part because technology shifts and mounting global competition hit so many companies hard, leading to extensive layoffs, and leaving some companies vulnerable to takeover. In early 1989, Wang Laboratories, which in the glory days of 1984 hired some 4,000 workers, began a series of job reductions that would over the next three years eliminate more than 17,000 employees. The state's unemployment rate began to edge up. In December, as the state government floundered in its attempts to address the quickly deteriorating financial situation, Standard & Poor's Corporation gave Massachusetts the lowest credit rating of the fifty states, and Moody's Investors Service followed soon after.

The outcry was deafening. Not surprisingly, in a state whose cynical populace is notorious for abandoning its sports heroes during a slump, Dukakis's popularity crashed amidst widespread criticism of his handling of the fiscal crisis; high-tech companies were excoriated for adopting shortsighted strategies; and many local critics forecast the imminent decline of Massachusetts. A 1989 survey funded by leading Massachusetts companies found that more than a quarter of the residents queried rated the state as either below average or as one of the worst states in which to live. Almost 50 percent of business leaders questioned about Massachusetts's business climate relative to other states rated it as below average or worst. Massachusetts had been hailed a few years before for building the economy of the future. Now it was condemned for running into an economic dead end that its high-technology businesses were not miraculous enough to find a way out of.

Many economists in 1989 predicted that the Massachusetts decline would be a short and modest one. Unfortunately, this was not the case. Over the next two years, bad news continued to accumulate at an alarming pace. By the end of 1991, the high-tech hiring boom seemed like a long-ago dream, and pink slips had become commonplace. In three years, the minicomputer mavens had hemorrhaged more than 40,000 jobs.

Industry analysts applauded the job cuts as part of a necessary downsizing and technology realignment. Data General and DEC had both yielded to competitive pressure to incorporate off-the-shelf parts in their new machines, and to take advantage of industry-standard software. Wang announced an agreement—unthinkable just a few years earlier—to sell the hardware of its longtime nemesis IBM under the Wang logo, focusing its own development efforts on software. But the strategic value of the downsizing was small comfort to the thousands of engineers, marketing staff, and factory workers who found themselves unemployed in an unforgiving market.

These and other high-tech job reductions created a chain reaction. Just as the technology buildup during the "Miracle" years had radiated out to create new jobs, so the high tech slide had a domino effect, knocking out jobs in finance, real estate, construction, materials, and other already weakened markets.

The state's unemployment statistics tell a grim story: By the end of 1990, Massachusetts in one eighteen-month period had gone from having the lowest unemployment rate among the eleven leading industrial states to having the highest. In June 1991, the 9.5 percent state jobless rate was well above the national average, and was in stark contrast to the low point of 2.7 percent in 1988. The "Help Wanted" section in the Sunday *Boston Globe* was reduced by more than half between 1987 and 1990. In a 1991 study released by Northeastern University's Center for Labor Market Studies, Massachusetts was found to have lost more than 275,000 jobs over a two-year period, more than any other state in the nation in more than thirty years.[37]

As many businesses scaled down or failed, and many workers searched for new jobs or saw their earning power dim, the once-torrid real estate market collapsed. According to *Banker and Tradesman,* a Boston publication, real estate foreclosures of homes and businesses statewide in the first six months of 1991 topped 13,000; in all of 1986, there had only been 1,198. The real estate bust also took its toll on regional financial institutions, many of which had invested heavily in real estate. Thirteen Massachusetts banks and credit unions failed in 1990, and another twenty failed in 1991, with the biggest shock waves set off by the collapse of the Bank of New England, the nation's fourth biggest bank failure at the time.

Not surprisingly, given the public's outraged reaction to the state's economic slide and fiscal crisis, Governor Dukakis opted not to seek reelection in 1990. In part because of a backlash against the state's Democratic administration and legislature, voters gave the nod to William Weld, the first Republican to be voted in to Massachusetts's top office in twenty years.

Under the more conservative Weld, the state intensified the cuts begun under Dukakis in a broad range of government programs, services, and local aid to cities and towns. During his campaign, Weld had promoted himself as a champion of business who would be devoted to economic development. But unlike Dukakis's often controversial efforts to nurture and spread the high-technology base, which included the creation of the Centers for Excellence and the Massachusetts Microelectronics Center, Weld's commitment to business was expected to take the form of more hands-off measures, such as tax cuts and business incentives.

By 1991, many of the programs originally hailed by the Dukakis administration as critical to the state's overall economic development had been largely dismantled. For example, Weld at first proposed eliminating all state funding for the Massachusetts Microelectronics Center, a semiconductor fabrication facility once touted as a model of what could be achieved when state government, industry, and academia combined forces. The Center eventually successfully fought to reinstate a little more than half its previous $4.5 million budget.

Far more alarming to many critics of the state's efforts to balance the budget under both Dukakis and Weld were continuing cuts in public higher education. From a high of $757 million in 1988, the state's higher education budget was dramatically reduced to $541 million for fiscal year 1992, resulting in canceled classes, layoffs, and an exodus of both students and faculty to institutions that appeared to have a brighter future. For a state that relied so heavily on the abilities of a well-educated work force, critics charged, such cuts not only were shortsighted, but also threatened the Commonwealth's very economic competitiveness.

Massachusetts was not suffering alone. The entire Northeast was beset by budget deficits, falling tax revenues, and rising unemployment. The national recession proved particularly punishing to states with strong stakes in high technology, defense, and financial services—all sectors that had enjoyed strong economic growth in the mid-1980s.

In Massachusetts, though, with the specter of the former "Miracle" hanging over it, things seemed worse. Indeed, the Massachusetts decline was worsened by the very prosperity that had preceded it. Many businesses did not want to expand, invest, or locate here in the mid- to late 1980s because costs had risen so high. By 1988 wages had risen to 108 percent of the national average, and in 1987 Boston had earned the dubious distinction of being the highest-priced housing market in the country.

Despite these hard times, not all the news was bad. Some of the excessive costs that had contributed to the downturn had already begun to correct themselves by the summer of 1991. Both wages and real estate prices fell as job opportunities dried up, and as houses, offices, and manufacturing space sat vacant. Venture capitalist Stephen Coit, who started an advocacy group in 1990 to bolster the credibility of the state's business community, claimed that the new bargain rates made the region ripe for business growth and expansion.

Moreover, even as bankruptcies soared, many high-tech companies in the 1990s were thriving as their particular products or industries experienced growth that—though starting from a smaller base—was just as explosive as that experienced by the minicomputer makers in the 1970s. Software, medical, and computer peripherals firms were among those flourishing in the midst of the recession. The boom years had ended, but innovation had not.

## BEYOND THE DECLINE

The so-called Massachusetts Miracle came to an inglorious close around 1989. During its peak, this economic boom had caught the attention not only of the nation, but of the entire world. When it ended, some observers dismissed the state's high-tech community as a flash in the pan.

But the miracle that most people contemplated was an illusion. Although it captured the world's attention in dramatic fashion, it ultimately did the region a disservice by putting the spotlight on a single, unusually explosive industry—computers. The extraordinary potential of this new industry focused attention on short-term monetary gains, causing people to overlook the long-term process that lay behind its creation. As we argued at the outset of this book, what should be labeled miraculous, and what is deservedly called the "Route 128 phenomenon," is not a short-lived local economic surge, but a stunning track record of innovation in a broad range of technologies, driven by the interactions of academia, industry, and the federal government.

Indeed, Massachusetts provides a special case study of the creative interplay among these three sectors due to their long and rich history of interactions in the state. The particular environment of the Boston area, combining industrial roots, a strong academic tradition, federal support, and an aptitude for risk taking and entrepreneurship, has made this region exceptionally fertile ground for the generation and growth of new ideas. The explosion of innovation attendant on the World War II effort, and the creation of Boston's high-tech complex, attest to this fertile union.

Understanding the significance of what has happened in the region—and what continues to happen—has a special urgency these days as the United States faces unprecedented competition in the world marketplace. As a fountainhead of new technology for well over a century, the Boston area is clearly a major resource. Moreover, America's growing preoccupation with its declining hegemony in world markets has not only sparked interest in nurturing the creativity found in such areas as Route 128, but also in trying to transfer these high-tech capabilities to other areas of the country.

It would be facile to suggest that the downturn in Massachusetts has had no impact. Both the pace of technology transfer and the rate of new company formation have been temporarily slowed. In 1984, during the peak of the "Miracle" years, entrepreneurs in the state founded 212 new high-tech companies, according to Mass Tech Times Incorporated. That figure had dwindled to 49 in 1988, and recent figures suggest the downward trend has continued. A combination of tighter capital and a forbidding economy has discouraged the easy flow of new technologies into the marketplace.

By all accounts, however, innovation still thrives in Massachusetts. The federal government is still sponsoring groundbreaking research in a broad array of fields, universities are still spinning off new ideas (MIT received more than 100 patents in 1990 alone), and entrepreneurs are still taking risks to bring their dreams to market.

Understanding the particular ingredients that have made Massachusetts a cradle of innovation will help this creativity and entrepreneurship continue to thrive. Moreover, the lessons of Boston's high-tech community may prove a vital resource as the nation prepares for the daunting economic challenges that lie ahead. The real miracle of Massachusetts is not dead. The challenge, now, is to make sure that it prospers.

# 6
# Beyond the "Miracle"

ACADEMIA, INDUSTRY, AND THE FEDERAL GOVERNMENT have been in continuous contact with one another since the passage of the Morrill Act in 1862. The give and take among these three sectors over the years has been instrumental not only in shaping the creation of Boston's high-tech community, but in fostering innovation nationwide.

The troika's interactions have been extraordinarily fruitful. They helped revolutionize agriculture; they led to the development of the best university system in the world; they helped propel the United States to world leadership in a wide variety of industries; they have sparked the creation of thousands of new companies and entirely new industries; they helped land the first person on the moon; and they helped create the most sophisticated defense system in the world. In the process, "high technology" has become a household term.

The Boston area has been in the thick of most of these developments. Throughout its long history, the region has been profoundly influenced by—and has had a profound influence on—the interactions of academia, government, and industry. This system of interaction has been in constant flux from the very start, adapting in response to challenges and changes that have arisen from economic pressures, national and international politics, social attitudes, and advances in science and technology. It has been a true evolution, proceeding haphazardly with little planning, and largely under the influence of personal visions and short-term concerns.

But this haphazard approach, which has extended to how people think about the region's accomplishments and needs, no longer seems appropriate. As the United States tries to marshal its resources in response to a new era of rapid change and global competition, it has become more important than ever for the nation to make the most of its strengths. As a fountainhead of new technology for well over a century, the Boston area is clearly a major resource. Yet, without a better understanding of how the region has evolved, and what it needs to continue to prosper, the nation is not in a position to capitalize on this resource.

In one sense, the importance of Boston's high-tech community has been well recognized. Regional planners both in the United States and abroad are striving to set up new high-tech communities in an effort to create jobs and spur economic growth. Many visitors to the Route 128 region hope to take away with them a blueprint for how to create or sustain a high-tech community.

All too often, however, they lose sight of what it took to create Boston's community, as well as distort what an area like this can be expected to produce. Replicating a region like Route 128 is not easily done. As this book has documented, it took a special blend of culture, people, institutions, and events to produce this hub of high technology. In addition, as the fallout from the Massachusetts Miracle demonstrates, high technology does not guarantee job growth or job stability or economic prosperity.

Nevertheless, there are valuable lessons to be learned from Boston's high-tech community. An analysis of Route 128's evolution suggests ways to preserve and enhance Boston's innovation-inducing infrastructure, as well as ways to foster the development of such centers of creativity in other areas. Moveover, these insights encourage a closer look at—and appreciation of—the special strengths not only of Massachusetts, but of the entire nation.

## LOOKING BACKWARD

Following are some broad observations on the process of innovation that has taken place in the Route 128 region in and around Boston.

- The Boston area's extensive research and educational infrastructure is the key to the region's ability to innovate.

Eastern Massachusetts is, above all else, a research and educational powerhouse. The extensive network of universities, hospitals, federal laboratories, and other research institutions draws some of the best and brightest students, faculty, and researchers from all over the world into the Boston area. The region's schools provide students with training in the latest technologies, and the network of research institutions and research-oriented firms gives many of them an opportunity to test their skills within the region. The principal output of this infrastructure is people with expertise and ideas.

High-tech companies are a side-effect of the talent that has been drawn to the region, and the diversity of research under way. Not surprisingly, many of these spin-off firms are strongly research oriented, taking advantage of the skills found locally, and doing business with the federal government, universities, and the research community. These companies may have no direct contact with the more lucrative—and visible—consumer market. Thus, high-tech companies in the region are, in many cases, an extension of the basic academic/research infrastructure. They provide a continuum of research-oriented organizations that help to ease new technologies into the high end of the marketplace.

It is also important to note that, in terms of economic impact, the nonprofit academic and federal research complex is itself big business in Massachusetts. Private higher education alone, in addition to attracting thousands of full-time students from out of state, is nearly a $10 billion industry within the state, and sustains almost 222,000 jobs, according to a 1991 study by the Association of Independent Colleges and Universities of Massachusetts.

- The Route 128 high-tech community was not planned, but evolved for specific reasons.

No one set out to create a high-tech mecca in the Boston area; it evolved out of the culture and, in particular, the academic and industrial traditions of the region. William Barton Rogers chose Boston as the site for his practical technological school, for example, because of its existing strengths in both education and industry. The strong interactions between MIT and industry over the next several decades, including consulting ties and the success of its graduates in industry, firmly established the Institute's culture of cooperation. And when President Roosevelt agreed to pump federal money into war research at the nation's universities, MIT, which had already established itself as a leader in the practically oriented research the government would need to support the war effort, garnered a large piece of the pie.

The region's evolution continued with the early blossoming of high-tech companies after World War II. As new and expanded markets for advanced technologies and expertise were opening up, the government was expanding its R&D support in areas ranging from microwaves to computers to guidance systems. The opportunities provided by the area's concentration of expertise in these fields, along with the practical New England culture and its pervasive, underlying entrepreneurial spirit, made the wealth of spin-offs that arose almost inevitable.

Finally, the "Miracle" years got a big boost from the commercial potential of one particular technology: minicomputers. By the early 1970s, computers had become inexpensive and versatile enough to win a vast marketplace. Computers, and the software that adapts them to specific uses, began to find application in virtually every field of business. This skyrocketing demand for computers and software helped to propel the Boston area into national prominence.

Once the high-tech community was established, the tendency of entrepreneurs to found spin-offs near the companies from which they originated also was crucial to the region's development. Again, there were concrete reasons for this trend. By staying in the greater Boston area, entrepreneurs could continue to tap into advances in their fields, and draw on a skilled labor pool, either from other firms or local universities. Moreover, many start-ups entered the market to fill a need they had identified nearby.

The critical mass of high-technology companies that has

resulted in the Route 128 region now largely sustains itself. In addition, it supports a wide variety of services, ranging from specialty machine shops and laboratories to venture capital and legal services, that make it much easier to start more firms nearby.

- State government has not played a major role in the development of the research infrastructure or the high-tech community.

As we have noted, the research infrastructure in Massachusetts evolved for very specific reasons. State government has played only a minimal role in the development of this infrastructure, and in the accompanying growth of high-tech companies. It could be argued that high tech flourished in Massachusetts in spite of state policy. During the 1970s and 1980s, for example, entrepreneurs continued to start up new firms there despite the widely held perception of a damaging antibusiness climate in the state.

The rise and fall of the Massachusetts Miracle occurred because of forces beyond the state's control. The unexpected explosion of the minicomputer and software markets in the 1970s that helped to thrust Massachusetts into the national spotlight was certainly not orchestrated by the state. Although politicians during the "Miracle" years tried to claim credit for the new technology-inspired prosperity, in fact, state policies had a limited impact on the region's development. Most state programs, including job training, capital availability, and tax policies, were implemented after the economic upturn began, and had only an incremental effect.

Nor was state government able significantly to soften the blows delivered by the fallout in the real estate market, the drop off in financial services, and the downscaling in the mature and flagging minicomputer industry in the late 1980s and early 1990s.

- Academia, the federal government, and industry have been motivated by fundamentally different goals.

Although these three sectors may share growing interests in many fields of science and engineering, their motivations have necessarily been different. Academia's role is to educate future generations of professionals, and to conduct intellectually motivated research as part of that process. The federal government has the broad welfare of society to address, including such areas as health, defense, and transportation. Industry must make a profit.

Each party thus has a different mission, and brings different perspectives to bear on research, the reasons for doing it, and how it should be pursued. Although these perspectives sometimes may be in agreement, they are often in conflict. Academia, for example, seeks to conduct research that has sufficient intellectual content to form the basis of publishable theses and scientific papers. In the name of academic integrity, it has often balked at too close an involvement with either industry or government. The federal government is typically more interested in applications, and may want to restrict access to research with military potential. Industry, for its part, tends to be interested in the direct profit potential of any research it conducts.

As a result, there has always been tension among the parties as they worry about having to compromise their needs, and seek to further their own interests. Such concerns have sometimes dampened or discouraged interaction, and have led to many attempts to define the "proper" role of each sector in any kind of interaction.

- Innovations have occurred when the "traditional" roles of these sectors and the boundaries between them have been challenged.

Looking back, it is remarkable to see how fruitful it can be when circumstances have forced—or allowed—the sectors to act and interact in new and different ways. Such tests of the status quo, large or small, have produced some surprising results.

The Morrill Act, for example, in providing federal support for education, and encouraging a more practical approach, not only redefined and expanded the role of the university in our society, but spurred the development of the profession of engineering. When MIT faculty ties to the chemical industry came under scrutiny, a department of chemical engineering was born. And it was a meeting of representatives of academica, industry, and local government in the mid-1930s that gave birth to the idea that led to the founding of the first formal high-tech venture capital firm.

Most striking of all, though, was the mobilization effort before and during World War II. When Vannevar Bush proposed that universities conduct wartime research, boundaries vanished, roles changed, and new roles appeared. With the incentive of responding to a national emergency, academia, government, and industry joined forces and proved extraordinarily productive.

Many start-up companies have been born of conflict as well. An employee at a university, a federal laboratory, or a company may have felt frustrated by the organization's lack of vision, for example, or its inability to offer the right opportunities for growth. This kind of frustration has often served as a stimulus for the students, researchers, or staff engineers who decided to test their ideas and expertise as entrepreneurs in the marketplace.

- Interactions among academia, government, and industry often produce unexpected results.

Because of tensions and differing perspectives among the sectors, as well as the unpredictable nature of research itself, the outcomes of interactions are likely to be different than expected. In some cases, such unexpected results have far eclipsed the original goals. Perhaps the most dramatic example was Jay Forrester's contract with the Navy to design a flight trainer. Although the Navy was understandably chagrined when Forrester's project ran way over budget, and became—in the Navy's mind—unnecessarily complex, the digital computer technology that resulted from this project-gone-awry not only became key to a far more important government "product"—a national air defense system—but also laid the groundwork for the birth of the minicomputer industry.

On the other hand, there have been many collaborations that have proven disappointing. During the 1920s, for example, MIT's formal Technology Plan to encourage companies to back university research never really got off the ground. Simultaneously, however, individual faculty members became so aggressive in pursuing outside consulting work that MIT's administration had to clamp down on the frequency of links that had formerly been championed as important, real-world ties.

- The economic benefits of the high-tech community and its underlying research/educational infrastructure are unpredictable and likely to fluctuate.

Because of its reliance on government funding, the economic performance of the research and development infrastructure of Eastern Massachusetts tends to be highly cyclical. As the government scaled back some of its research efforts with the winding down of the Apollo project and the end of the Vietnam War, for example, many Massachusetts engineers and scientists found themselves without jobs. The rise of the Strategic Defense Initiative in the 1980s, on the other hand, gave the infrastructure a boost.

Perhaps an even more significant variable is the unpredictable nature of emerging technologies. The explosion of the computer and software industries in the 1970s, for example, took the area by storm, even diversifying the state's industrial base. In more than a century of high-tech innovation the region had never seen a technology with such widespread potential. In truth, it is unlikely that innovations with such a profound impact will be emerging frequently in the future. Predictions of similarly explosive growth in the biotech industry, for example, have yet to materialize. As a result, the region probably will not experience anything like the cataclysmic changes of the 1970s and early 1980s again soon. However, the high-tech sector will continue to generate new jobs at a modest but steady pace as new firms start up, and as the existing diverse array of technology companies expands.

The high-tech companies that have the greatest impact on the region's economy, not surprisingly, are the large, established companies, organizations that typically are no longer in the start-up phase, but are part of a mature industry. These companies, such as the minicomputer makers and defense contractors, are subject to a different mix of competitive pressures and concerns than are start-ups.

- The ideas and expertise from this area often benefit the nation more than the region.

The existence of the research/educational infrastructure and the high-tech community in Massachusetts does not guarantee that the state will benefit directly from all the innovations that result. Although Bell Telephone had its beginnings in Boston, it relocated to New York at the turn of the century before becoming a huge corporation employing thousands. MIT's collaborations with the oil industry that began in the 1920s had no impact on the region at all. The magnetic core memory developed by An Wang primarily benefited another out-of-state company: IBM. And many of the radar systems developed by Lincoln Lab and MITRE were manufactured by companies based elsewhere, such as Westinghouse.

Clearly, the research infrastructure, with its broad connections, is a national resource at least as much as it is a state resource. Although innovations may originate in Massachusetts, they often reach the product or profit stage in other parts of the country. In addition, the alumni of area schools are a national resource. Although a large percentage of graduates do elect to stay in the region, the majority carry their expertise out of state.

- Innovations at all levels have been driven by individuals, not organizations.

Innovation relies critically on the persistence and creativity of individuals acting on their personal visions. The ability of individuals to take risks and challenge the status quo has led to the greatest advances both in technology and in how the different sectors interact.

A patient but driven William Barton Rogers, for example,

finally founded MIT in 1861 as a school that merged the academic and industrial traditions of New England. Vannevar Bush made an end run around the governmental bureaucracy and revolutionized higher education by bringing federally sponsored research to campus. And Karl Taylor Compton and a handful of associates from academia, government, and industry pushed through a formal venture capital firm to support new technology ventures.

Even within the research community, the overwhelming majority of interactions proceed at the individual level. At universities, individual faculty typically must seek funding for their own research projects from federal agencies or industry, and academic theses are meant to demonstrate the degree candidate's ability to make original contributions to the course of research.

Above all, it has been individual entrepreneurs who drove innovation by starting new companies, by introducing new technologies to the marketplace, and by attempting to make something that has never been made before. From Ken Olsen of giant DEC to the founders of the latest software operation, these are the people who have given form to the theoretical knowledge of MIT and the rest of the region's prolific research/industrial complex.

## LOOKING FORWARD

The story of Boston's high-tech community illustrates how fruitful the interactions among academia, the federal government, and industry can be. Not surprisingly, the collaborations between these three sectors have been especially effective during times of national need. During World War II, in particular, academia, government, and industry were able to form unusual liaisons that were vital to the nation's successful mobilization, and that had the unintended effect of sparking the astonishing growth of Boston's high-tech community.

Now the United States is facing a new kind of national emergency, an economic one. For the first time since academia, industry, and the federal government came into formal contact, the country's economic influence is not growing, it is declining. In most key marketplaces, global competition is the rule, not the exception, and Europe and Japan now compete as equals with the United States. Accordingly, the pressure is on as never before to boost the nation's productivity and competitiveness.

Technology, in particular, will continue to be one of the central drivers of change. As the pace of technological advances accelerates, and as the rush to commercialize new technologies picks up, it will be more important than ever to maintain our position at the leading edge. The Route 128 region—a fountainhead of new technology for well over a century—is clearly an important national resource in this regard. Yet it is unclear how quickly and how well the region and the nation as a whole will be able to respond to this kind of economic emergency. High technology is just one piece of the puzzle. The United States is struggling with a broad range of strategic issues, including boosting its manufacturing competence, analyzing the value of governmental/industrial liaisons, adopting new management techniques, and so on.

Unquestionably, the United States does need to make changes. The ability to change and adapt will be a critical precursor of success in the decades to come. Many of the management and manufacturing techniques being adopted from Japanese industry, for example, including concepts such as team work and quality control, are filling a long-felt need in American industry. In its eagerness to become more competitive, however, the United States must be careful not to lose sight of what it does best. In the area of innovation, for example, this country is still unexcelled. We believe that the qualities that have made America so creative should be harnessed not only in ensuring that that creativity continues, but in addressing other areas where the United States has lagged behind.

The following list of suggestions, based on the insights that the Boston area's experience provides, can help the region and the country make the most of their creative potential.

- Preserve and strengthen the research/educational infrastructure in the United States.

The United States' creativity and leadership in basic research rest directly on the excellence of the country's educational and research institutions. At present, the American university system is recognized as the best in the world. It is imperative that this resource be nourished and preserved. Most important, these institutions educate the nation's professional work force, including scientists, engineers, medical doctors, and managers. These are the people who will continue to make research advances, refine them, and put them to work in government and industry.

To do so in a world of mounting international competition, these graduates, as well as their teachers, will need a global perspective. The ties to the research and business communities of other nations that have long been a part of the academic community have thus become more critical than ever. These links should be nurtured and expanded to help keep students from becoming insular or isolated, as well as to keep them abreast of the best practices around the world in everything from science to management.

It is also crucial to continue a broad agenda of basic research at universities. Although most incremental industrial innovations occur in industry, the basic research that makes them possible has been the special province of universities. Furthermore, the diversity of this research is important. Some of the most significant advances of this century have come either from unexpected quarters, or from an unusual cross-fertilization of perspectives and disciplines.

Higher education, however, does not stand alone. Even the best universities in the world cannot maintain their excellence without a continuing stream of well-prepared entering students. For the United States to make the best use of this system, it must maintain the quality of the primary and secondary schools that feed it. Strengthening education in the United States has the added benefit of raising the general quality of the nation's work force. In a world where both manufacturing and services are becoming increasingly sophisticated, and where products themselves are more complex, well-educated labor at all levels becomes an essential component of economic success.

- Recognize the special roles of Route 128 and other similar regions.

There are very few regions in the United States that not only have a top-notch infrastructure for education and research but also a large-scale and vital high-tech entrepreneurial community that encourages the movement of advanced ideas and expertise into the marketplace. The Boston area, Silicon Valley, and Southern California are by far the leaders in this special category. In each of these areas, a particular regional culture has emerged that has encouraged high-tech entrepreneurship and created a critical mass that helps sustain creativity. Because of their competitive but stimulating environments—as well as the presence of local markets, local talent, and specialized support services—these areas have become among the few accepted "centers" for start-up companies.

These distinctive cultural qualities and economies of scale are extremely difficult to duplicate. There are many excellent academic centers around the world, but there are few Route 128s or Silicon Valleys. In the interest of bolstering their economies, many regions have tried various schemes, including research parks and incubators, to encourage the commercialization of new technologies emerging from local universities, but to little avail. Companies will almost always start up where there are compelling reasons for them to do so.

It is important to recognize the special nature of areas like Route 128, both as centers of research and education and as centers of high-tech entrepreneurship. Because of the rich environments they offer, they can provide particularly valuable places for both industry and government to locate advanced technology research facilities, which in turn further encourage the passage of new technologies into the marketplace.

- The state of Massachusetts should work to capitalize on its strengths.

Although Massachusetts did relatively little to cultivate the development of the high-tech community in the Boston area, it can help maintain and enhance it. Recognizing the importance of the underlying educational/research infrastructure in drawing talented people, research dollars, and jobs into the region, for example, the state should take steps to reinforce its role as an educational center, stepping up its support for higher education, and encouraging strong primary and secondary schools. In addition, the state should more actively promote the strengths and record of achievement of its extended educational and research network. Since the federal government funds the vast majority of research and development in the region, it must be kept aware of the benefits derived from that funding.

Moreover, Massachusetts should aggressively market the state to industry as a research center. As a fountainhead of new technology, the Boston area is a logical place to locate leading-edge research facilities in a broad array of fields, including computers, software, artificial intelligence, biotechnology, electronics, and materials. Like the organizations already on and around Route 128, these facilities would be able to tap into the infrastructure and strengthen it at the same time.

Finally, the state could become a kind of mediator, bringing key representatives of academia, industry, and the federal government together to discuss mutual interests and goals pertinent to the region. For the most part, there have been few formal connections among these sectors; even industry and academia, with a number of seemingly overlapping interests, are only loosely allied.

- Other regions should build on their own particular strengths.

Massachusetts's distinctive regional strength is education and research—a strength that led to the creation of the high-tech community. This is not a realistic niche for every area. Nor can any region or nation count on high-tech companies to spring up and provide a major source of new jobs. Most emerging technologies, even if successful, do not have that kind of dramatic potential by themselves.

Although building a full-scale Route 128 somewhere else

may not be a reasonable goal, research and high technology can be valuable components of regional planning, if they are considered within the context of an area's existing or potential strengths. Any programs for new technology development should consider established industries, the skills of the labor force, educational institutions, and location, for example. Such a self-assessment, conducted jointly by the state, the academic community, and local industry, makes possible a realistic plan for drawing the appropriate people, companies, or research facilities to the area.

• Don't overmanage the "system."

In the fever to "catch up" with foreign competitors in the world marketplace, some leaders are suggesting that the United States should channel its resources into specific selected areas of technology. At the same time, politicians anxious to draw federal research funds into their districts are sidestepping the established peer review procedures typically used to determine locations for research facilities or programs. These trends threaten to cut into the effectiveness of the nation's system of basic research.

One of the key lessons from Boston's high-tech community is that it pays to give creativity a loose rein. For the most part, the federal government has played a background role in funding research, both in universities and to some extent in its own laboratories. The purpose of this approach is to explore a number of promising avenues and alternatives and to strengthen the basic understanding of a wide variety of fields. Although the government funds the vast majority of research, and may develop certain targeted technologies for military use, it has rarely stepped in to push a specific technology to the commercial marketplace.

It is dangerous for the federal government to try to outguess the marketplace or the course of research. It makes more sense to rely on the creativity and entrepreneurship inherent in the system to adapt to meet these needs and demands. Even during World War II, despite the strong sense of mission and the need for concrete results, government did not guide university research with a heavy hand. Similarly, it does not make sense to play politics with decisions on where to site research. A new research facility may win prestige for an isolated school or win votes for a politician, but much of its potential may be wasted in the wrong locale.

- Celebrate the differences among academia, industry, and the federal government.

Although they may clash at times, in the long run, the differing missions and perspectives of academia, industry, and the federal government are complementary and mutually enriching. By any measure, these sectors have not worked smoothly and reliably together like parts of a well-oiled machine. Instead, they have pushed and pulled at one another over the years in what has proven to be a creative tug of war.

On the one hand, the different missions of the three sectors have meant that the United States never has achieved—and probably never will achieve—the kinds of cooperative partnerships that exist, for example, in Japan. On the other hand, the differences among the forces—though standing in the way of a long-lasting partnership—lie at the heart of the productivity and adaptability of the country.

As the experience of the Boston area illustrates, the perspectives of academia, of industry, and of the federal government have all grown up together, actively molding and shaping one another in response to a rapidly changing world. As the pace of change increases on all fronts—technological, political, and economic—it will be crucial for each of these sectors to preserve its own missions while listening to the needs of the others, and seeking new balances.

- Encourage experimentation.

The key to adjusting to change in the world is to encourage experimentation. As we have noted, real progress has been made when barriers are breached and traditional beliefs are challenged. More than ever, it is important for academia, industry, and the federal government to be bold in striking up and testing new relationships within the context of their basic missions.

These relationships may take many forms. Academia and industry, for example, must find new ways to bridge the gap between them as technology becomes an increasingly powerful driving force in the marketplace. Both universities and industry should expand their networks of international relationships, in order to prepare for a more global economy. And academia, industry, and the federal government must all lay the groundwork that will encourage the actions and creativity of individuals.

## AN AMERICAN MIRACLE

The Route 128 region is at the same time both less and more than people think it is. It is not, for example, a fountainhead that steadily spews forth thousands of new high-tech jobs. It does not create hundreds, or even dozens, of huge new companies. And the health of the large companies that have been created here is not a measure of the real value of what is happening in the high-tech community.

On the other hand, the forces at work in this region do have a profound impact on the region, the state, and the nation. The effects are much more subtle, long-term, and far-reaching, however, than the creation of a few companies here and there. The Boston area is a cradle of innovation. It represents, in microcosm, the extraordinarily complex process by which a broad spectrum of new ideas and new technologies are created, developed, refined, applied, and brought to the marketplace in our society.

This process has developed for over a century, changing and growing under the influence of the times and the people who have been a part of it. Furthermore, this evolution has been so gradual, and so intertwined with events both within the region and from afar, that it has been difficult to grasp its significance, or even—sometimes—its existence.

In the United States, academia, the federal government, and industry have been the mind, muscle, and hand of innovation in high technology. The academic community trains the minds of the diverse people who will become the innovators, and generates many of the ideas that go on to become applied processes or products. The federal government uses its muscle both to push and to pull the innovative process, funding high-tech research at universities, and providing a range of interim markets for technological expertise. Industry puts the ideas and expertise to work.

Each of these sectors has uniquely American characteristics. The American system of government, for example, is founded on principles that celebrate the individual and encourage initiative, resisting unnecessary intervention into personal and public affairs. The business community embraces innovation, and has raised risk-taking entrepreneurs to the popular status of folk heroes, largely irrespective of their success or failure. The system of higher education, which puts a premium on diversity of opinion and research, is unexcelled.

A culture that stresses such individuality is inherently less stable and less harmonious, and some people have argued that these traits will hurt the United States in the global marketplace, making it harder for the country to challenge more stable or homogeneous foreign competitors. A typical American, after all, is someone who challenges authority, freely airs competing views, and wants to make a personal mark on the world. Indeed, this individualistic and entrepreneurial attitude is one contributor to job instability in America. Economist David Birch has pointed out that over half of all jobs in the United States are replaced every five years as businesses start up, fail, and start again.

Such turbulence may be unsettling at times, but these are the same traits that make the nation so innovative. The Boston high-tech community's evolution has depended on individuals and institutions that have made the most of these fundamentally American characteristics. These qualities not only underlie the nation's past successes, they are its hope for the future. A culture that looks for and embraces the new, and that is willing to take chances and rise to new challenges, is uniquely suited to adapting to the demands of a rapidly changing world.

We are at our best when we are most American. The innovation that has flourished in the Route 128 region is not so much a Massachusetts miracle as an American miracle. What has evolved in the greater Boston area is a system that builds on our nation's inherent strengths, its diversity, its Yankee ingenuity, and its strongly entrepreneurial culture, to generate a steady stream of new ideas and technologies, to nurture them, and to ease them into the marketplace.

The qualities of risk taking, entrepreneurship, ingenuity, and diversity that built the Boston area into a high-tech mecca are the same qualities that the country must marshal as it faces the considerable economic challenges that lie ahead. Although we can do a better job of learning from other countries as we go toe-to-toe in the marketplace, we should not forget our own innate strengths. We must be ourselves. This is the real lesson of Boston's high-tech community.

# Notes

## CHAPTER 1

1. Authors' interview with Kenneth Olsen, Maynard, Massachusetts, 12 November 1985.
2. Authors' interview with Mitchell Kapor, Cambridge, Massachusetts, 6 November 1985.

## CHAPTER 2

1. James Botkin, Dan Dimancescu, and Ray Stata, *Global Stakes: The Future of High Technology in America* (Cambridge, Mass.: Ballinger Publishing Company, 1982).
2. Lawrence A. Cremin, *American Education: The Colonial Experience 1607–1783* (New York: Harper & Row, 1970).
3. James M. McPherson, *Battle Cry of Freedom: The Civil War Era* (New York: Oxford University Press, 1988).
4. David F. Noble, *America By Design: Science, Technology, and the Rise of Corporate Capitalism* (Oxford: Oxford University Press, 1977).
5. Daniel J. Boorstin, *The Americans: The National Experience* (New York: Vintage Books, 1965).

6. L. S. Bryant and J. B. Rae (eds.), *Lowell: An Early American Industrial Community* (Cambridge, Mass.: The Technology Press, 1950).
7. Samuel C. Prescott, *When M.I.T. Was "Boston Tech"* (Cambridge, Mass.: The Technology Press, 1954).
8. Ibid.
9. E. J. Kahn, Jr., *The Problem Solvers: A History of Arthur D. Little, Inc.* (Boston: Little, Brown and Company, 1986).
10. Ibid.
11. Ibid.
12. David A. Hounshell and John Kenly Smith, Jr., "The Nylon Drama," *American Heritage of Invention & Technology,* Fall 1988.
13. George Wise, *Willis R. Whitney, General Electric, and the Origins of U.S. Industrial Research* (New York: Columbia University Press, 1985).
14. Hounshell and Smith, "The Nylon Drama."
15. Francis A. Walker, "The Place of the Schools of Technology in American Education," *Educational Review,* October 1891.
16. Noble, *America By Design.*
17. Ibid.
18. Hoyt C. Hottel, "Chemical Engineering at MIT in the '20s" (speech delivered at the MIT Centennial of Chemical Engineering), 7 October 1988.
19. William H. Walker, "Chemical Research and Industrial Progress," *Scientific American Supplement,* 1 July 1911.
20. F. Leroy Foster, *Sponsored Research at M.I.T., Vol. I: 1900–1940* (published internally at MIT) 1984.
21. John W. Servos, "The Industrial Relations of Science: Chemical Engineering at MIT, 1900–1939," *ISIS,* 1980.
22. Authors' interview with Charles Adams, Lexington, Massachusetts, 19 September 1985.
23. Otto J. Scott, *The Creative Ordeal* (New York: Atheneum, 1974).
24. James R. Killian, Jr., *The Education of a College President: A Memoir* (Cambridge, Mass.: The MIT Press, 1985).

25. Martin Kenney, *Biotechnology: The University–Industrial Complex* (New Haven: Yale University Press, 1986).

## CHAPTER 3

1. Vannevar Bush, *Pieces of the Action* (New York: William Morrow and Company, Inc., 1970).
2. Ibid.
3. Authors' interview with Howard Whitman, Wilmington, Massachusetts, October 1988.
4. Vannevar Bush, *Science—The Endless Frontier: A Report to the President on a Program for Postwar Scientific Research,* July 1945 (reprinted by the National Science Foundation, May 1980).
5. Ibid.
6. Ibid.

## CHAPTER 4

1. Joint Board for the Metropolitan Master Highway Plan, *Master Highway Plan,* 1948.
2. "New England Highway Upsets Old Way of Life," *Business Week,* 14 May 1955.
3. Authors' interview with George Hatsopoulos, Waltham, Massachusetts, 18 November 1985.
4. Authors' interview with John Volpe, Nahant, Massachusetts, February 1988.
5. Authors' interview with Douglas Ross, Waltham, Massachusetts, 23 September 1985.
6. Authors' interview with Kenneth Olsen, Maynard, Massachusetts, 12 November 1985.
7. Authors' interview with William Congleton, Woburn, Massachusetts, July 1988.

8. Authors' interview with Denis Robinson, Burlington, Massachusetts, 30 October 1985.
9. Authors' interview with William Congleton.
10. Authors' interview with Peter Brooke, Boston, Massachusetts, July 1988.
11. An Wang, *Lessons: An Autobiography* (Reading, Massachusetts: Addison-Wesley, 1986).
12. Authors' interview with Howard Whitman, Wilmington, Massachusetts, October 1988.
13. Authors' interview with Jack Bush, Bedford, Massachusetts, 11 December 1985.
14. Authors' interview with Dana Atchley, Burlington, Massachusetts, 3 October 1985.
15. Authors' interview with William Brown, Boston, Massachusetts, July 1988.
16. Authors' interview with Peter Brooke.
17. Authors' interview with Jane Morris, Wellesley, Massachusetts, August 1988.
18. Authors' interview with Peter Brooke.
19. Authors' interview with Richard Farrell, Boston, Massachusetts, July 1988.
20. Authors' interview with Edson de Castro, Westboro, Massachusetts, 10 October 1985.
21. Authors' interview with Alexander d'Arbeloff, Boston, Massachusetts, 13 September 1985.
22. Authors' interview with Douglas Ross.
23. Authors' interview with William Foster, Marlboro, Massachusetts, 7 October 1985.
24. Authors' interview with Robert P. Berkowitz, Framingham, Massachusetts, August 1988.
25. Authors' interview with William Poduska, Chelmsford, Massachusetts, 24 October 1985.
26. Authors' interview with Raymond Kurzweil, Waltham, Massachusetts, 12 February 1986.

27. David Birch, *Job Creation in America: How Our Smallest Companies Put the Most People to Work* (New York: The Free Press, 1987).
28. Authors' interview with Peter Brooke.
29. John W. Wilson, *The New Venturers: Inside the High-Stakes World of Venture Capital* (Reading, Mass.: Addison-Wesley, 1985).
30. Authors' interview with Kenneth Fisher, Marlboro, Massachusetts, 19 November 1985.
31. Authors' interview with William Egan, Boston, Massachusetts, July 1988.
32. Authors' interview with Roland Tibbetts, Washington, D.C., 25 January 1989.
33. Authors' interview with Robert Weiss, Andover, Massachusetts, January 1989.
34. Authors' interview with Ray Stata, Norwood, Massachusetts, 12 September 1985.
35. Authors' interview with William Poduska.
36. Authors' interview with Kenneth Fisher.

## CHAPTER 5

1. Authors' interview with Kenneth Olsen, Maynard, Massachusetts, 12 November 1985.
2. Unpublished study that accompanied the release of the MIT commemorative publication, "Event 128: A Salute to Founders," 29 April 1989.
3. Authors' interview with Amar Bose, Framingham, Massachusetts, 16 October 1985.
4. Authors' interview with George Hatsopoulos, Waltham, Massachusetts, 18 November 1985.
5. Authors' interview with Alexander d'Arbeloff, Boston, Massachusetts, 13 September 1985.
6. Authors' interview with Kenneth Fisher, Marlboro, Massachusetts, 19 November 1985.

7. Authors' interview with Thomas Fraser, Cambridge, Massachusetts, 17 October 1985.
8. Authors' interview with Mitchell Kapor, Cambridge, Massachusetts, 6 November 1985.
9. Authors' interview with William Foster, Marlboro, Massachusetts, 7 October 1985.
10. Authors' interview with Philippe Villers, Burlington, Massachusetts, 25 September 1985.
11. Authors' interview with Ronald Gruner, Acton, Massachusetts, 22 January 1986.
12. Steven Levy, *Hackers: Heroes of the Computer Revolution* (New York: Dell Publishing Company, 1984).
13. Authors' interview with Russell Noftsker, Cambridge, Massachusetts, 18 February 1986.
14. Authors' interview with Richard Greenblatt, Cambridge, Massachusetts, 18 September 1985.
15. Authors' interview with Peter Gabel, Concord, Massachusetts, February 1989.
16. "Business in Isotopes," *Fortune,* December 1947.
17. Christopher L. Taylor, *New Enterprises Descended from a Technically Based Company* (unpublished S. M. thesis, MIT Sloan School of Management, 1981).
18. Glenn Rifkin and George Harrar, *The Ultimate Entrepreneur* (Chicago: Contemporary Books, 1988).
19. Authors' interview with Edson de Castro, Westboro, Massachusetts, 10 October 1985.
20. David Wessel, "Data General's Progeny," *The Boston Globe,* 6 September 1983.
21. Unpublished study that accompanied the release of the MIT commemorative publication, "Event 128: A Salute to Founders," 29 April 1989.
22. Authors' interview with William Foster.
23. Authors' interview with Mitchell Kapor.
24. Authors' interview with William Bowman, Cambridge, Massachusetts, 7 November 1985.

25. Authors' interview with Philippe Villers.
26. Authors' interview with William Foster.
27. Authors' interview with Mitchell Kapor.
28. Authors' interview with Kenneth Fisher.
29. Authors' interview with Frank Morgan, Boston, Massachusetts, August 1988.
30. Authors' interview with Aaron Kleiner, Waltham, Massachusetts, 12 February 1986.
31. Authors' interview with David Blohm, Cambridge, Massachusetts, February 1989.
32. Authors' interview with Stephen Coit, Waltham, Massachusetts, August 1988.
33. Authors' interview with Richard Farrell, Boston, Massachusetts, July 1988.
34. Authors' interview with William Egan, Boston, Massachusetts, July 1988.
35. Authors' interview with Stuart Madnick, Cambridge, Massachusetts, August 1988.
36. Authors' interview with William Egan.
37. Paul E. Harrington and Andrew Sum, "The Economic Recession in Massachusetts: Its Impacts on Labor Markets and Workers Through June 1991" (report by the Center for Labor Market Studies, Northeastern University), 8 July 1991.

# Bibliography

BOOKS

ADAMS, RUSSELL B., JR. *The Boston Money Tree.* New York: Thomas Y. Crowell Company, 1977.

BARBEAU, JOSEPH E. *Second to None: Seventy-five Years of Leadership in the Cooperative Education Movement.* Boston: Custom Book Program of Northeastern University, 1985.

BIRCH, DAVID. *Job Creation in America: How Our Smallest Companies Put the Most People to Work.* New York: The Free Press, 1987.

BOK, DEREK. *Universities and the Future of America.* Durham, N.C.: Duke University Press, 1990.

BOORSTIN, DANIEL J. *The Americans: The Colonial Experience.* New York: Vintage Books, 1958.

BOORSTIN, DANIEL J. *The Americans: The Democratic Experience.* New York: Vintage Books, 1973.

BOORSTIN, DANIEL J. *The Americans: The National Experience.* New York: Vintage Books, 1965.

BOTKIN, JAMES, DAN DIMANCESCU, and RAY STATA. *Global Stakes: The Future of High Technology in America.* Cambridge: Ballinger Publishing Company, 1982.

BUSH, VANNEVAR. *Pieces of the Action.* New York: William Morrow and Company, 1970.

BYLINSKI, GENE. *The Innovation Millionaires: How They Succeed.* New York: Charles Scribner's Sons, 1976.

CHAMBERLAIN, JOHN. *The Enterprising Americans: A Business History of the United States.* New York: Harper & Row, 1963.

CREMIN, LAWRENCE A. *American Education: The Colonial Experience 1607–1783.* New York: Harper & Row, 1970.

DERTOUZOS, MICHAEL L., et al. *Made in America: Regaining the Productive Edge.* Cambridge, Mass.: The MIT Press, 1989.

DUPREE, A. HUNTER. *Science in the Federal Government: A History of Policies and Activities to 1940.* New York: Harper & Row, 1957.

FLORIDA, RICHARD, and MARTIN KENNEY. *The Breakthrough Illusion: Corporate America's Failure to Move from Innovation to Mass Production.* New York: Basic Books, 1990.

FOSTER, RICHARD N. *Innovation: The Attacker's Advantage.* New York: Summit Books, 1986.

HALL, PETER, and ANN MARKUSEN, eds. *Silicon Landscapes.* Boston: Allen & Unwin, 1985.

HOY, JOHN C., and MELVIN H. BERNSTEIN, eds. *Business and Academia: Partners in New England's Economic Renewal.* Hanover, N.H.: University Press of New England, 1981.

HOY, JOHN C., and MELVIN H. BERNSTEIN, eds. *Financing Higher Education: The Public Investment.* Boston: Auburn House Publishing Company, 1982.

HOY, JOHN C., and MELVIN H. BERNSTEIN, eds. *New England's Vital Resource: The Labor Force.* Washington, D.C.: American Council on Education, 1982.

KAHN, E. J., JR. *The Problem Solvers: A History of Arthur D. Little, Inc.* Boston: Little, Brown and Company, 1986.

KENNEY, MARTIN. *Biotechnology: The University–Industrial Complex.* New Haven: Yale University Press, 1986.

KIDDER, TRACY. *The Soul of a New Machine.* Boston: Little, Brown and Company, 1981.

KILLIAN, JAMES R., JR. *The Education of a College President: A Memoir.* Cambridge: The MIT Press, 1985.

LAMPE, DAVID R., ed. *The Massachusetts Miracle: High Technology and Economic Revitalization.* Cambridge, Mass.: The MIT Press, 1988.

LEVENTMAN, PAULA Q. *Professionals Out of Work.* New York: The Free Press, 1981.

LEVY, STEVEN. *Hackers: Heroes of the Computer Revolution.* New York: Dell Publishing Company, 1984.

MACLAURIN, W. RUPERT. *Invention & Innovation in the Radio Industry.* New York: The Macmillan Company, 1949.

MALONE, MICHAEL S. *The Big Score: The Billion-Dollar Story of Silicon Valley.* Garden City, N.Y.: Doubleday & Company, 1985.

MCPHERSON, JAMES M. *Battle Cry of Freedom: The Civil War Era.* New York: Oxford University Press, 1988.

MENDELSOHN, EVERETT, MERRITT ROE SMITH, and PETER WEINGART, eds. *Science, Technology, and the Military.* Vol. 2. Dordrecht: Kluwer Academic Publishers, 1988.

MORISON, ELTING E. *Men, Machines, and Modern Times.* Cambridge, Mass.: The MIT Press, 1966.

MORISON, SAMUEL ELIOT. *Builders of the Bay Colony.* Boston: Houghton Mifflin Company, 1930.

MOWERY, DAVID C. *Technology and the Pursuit of Economic Growth.* Cambridge: Cambridge University Press, 1989.

NOBLE, DAVID F. *America by Design: Science, Technology, and the Rise of Corporate Capitalism.* Oxford: Oxford University Press, 1977.

NOONE, CHARLES M., and STANLEY M. RUBEL. *SBICs: Pioneers in Organized Venture Capital.* Chicago: Capital Publishing Company, 1970.

ORTH, CHARLES D., III, JOSEPH C. BAILEY, and FRANCIS W. WOLEK. *Administering Research and Development.* Homewood, Ill.: Richard D. Irwin Inc., and The Dorsey Press, 1964.

OSBORNE, DAVID. *Laboratories of Democracy.* Boston: Harvard Business School Press, 1988.

PRESCOTT, SAMUEL C. *When M.I.T. Was "Boston Tech."* Cambridge: The Technology Press, 1954.

PUGH, EMERSON W. *Memories That Shaped an Industry.* Cambridge, Mass.: The MIT Press, 1984.

PURSELL, CARROLL W., JR., ed. *Technology in America: A History of Individuals and Ideas.* Cambridge, Mass.: The MIT Press, 1981.

RAND, CHRISTOPHER. *Cambridge, USA: Hub of a New World.* New York: Oxford University Press, 1964.

REDMOND, KENT C., and THOMAS M. SMITH. *Project Whirlwind: The History of a Pioneer Computer.* Bedford, Mass.: Digital Press, 1980.

REES, JOHN, ed. *Technology, Regions, and Policy.* Totowa, N.J.: Rowman & Littlefield, 1986.

RIFKIN, GLENN, and GEORGE HARRAR. *The Ultimate Entrepreneur.* Chicago: Contemporary Books, 1988.

ROBERTS, EDWARD B. *Entrepreneurs in High Technology: Lessons from MIT and Beyond.* New York: Oxford University Press, 1991.

ROGERS, EVERETT M., and JUDITH K. LARSEN. *Silicon Valley Fever.* New York: Basic Books, 1984.

RUDOLPH, FREDERICK. *The American College and University.* New York: Alfred A. Knopf, 1962.

SCOTT, OTTO J. *The Creative Ordeal: The Story of Raytheon.* New York: Atheneum, 1974.

SLATER, ROBERT. *Portraits in Silicon.* Cambridge, Mass.: The MIT Press, 1987.

SMITH, MERRITT ROE, ed. *Military Enterprise and Technological Change: Perspectives on the American Experience.* Cambridge, Mass.: The MIT Press, 1985.

SMITH, RICHARD NORTON. *The Harvard Century: The Making of a University to a Nation.* New York: Simon & Schuster, 1986.

STONE, ORRA L. *History of Massachusetts Industries: Their Inception, Growth and Success.* Vol. I. Boston: The S. J. Clarke Publishing Company, 1930.

VESPER, KARL H. *New Venture Strategies.* Englewood Cliffs, N.J.: Prentice-Hall, 1980.

WANG, AN. *Lessons: An Autobiography.* Reading, Mass.: Addison-Wesley, 1986.

WARNER, SAM B., JR. *Streetcar Suburbs: The Process of Growth in Boston 1870–1900.* New York: Atheneum, 1969.

WENSBERG, PETER C. *Land's Polaroid.* Boston: Houghton Mifflin Company, 1987.

WICKSTEED, SEGAL QUINCE. *The Cambridge Phenomenon: The Growth of High Technology Industry in a University Town.* London: Brand Brothers and Company, 1985.

WILDES, KARL L., and NILO A. LINDGREN. *A Century of Electrical Engineering and Computer Science at MIT, 1882–1982.* Cambridge: The MIT Press, 1985.

WILLIAMS, BEN AMES, JR. *Bank of Boston 200: A History of New England's Leading Bank, 1784–1984.* Boston: Houghton Mifflin Company, 1984.

WILSON, JOHN W. *The New Venturers.* Reading, Mass.: Addison-Wesley Publishing Company, 1985.

WISE, GEORGE. *Willis R. Whitney, General Electric, and the Origins of U.S. Industrial Research.* New York: Columbia University Press, 1985.

WYLIE, FRANCIS W. *MIT in Perspective.* Boston: Little, Brown and Company, 1975.

## MAGAZINE ARTICLES

BAGAMERY, ANNE. " 'No Policy Is Good Policy.' " *Forbes,* June 18, 1984.

BEATTY, JACK. "Lowell Weaves a Spell." *National Geographic Traveler,* 2 (Autumn 1985).

BELLO, FRANCIS. "The Prudent Boston Gamble." *Fortune,* November 1952.

BENNER, SUSAN. "Storm Clouds Over Silicon Valley." *Inc.,* September 1982.

BLUMENTHAL, DAVID, et al. "Industrial Support of University Research in Biotechnology." *Science,* 231 (January 17, 1986).

BLUMENTHAL, DAVID, et al. "University-Industry Research Relationships in Biotechnology: Implications for the University." *Science,* 232 (June 13, 1986).

BOLLINGER, LYNN, and JAMES M. UTTERBACK. "A Review of the Literature and Hypotheses on New Technology-Based Firms." *Research Policy,* 12 (1983).

BRABEN, DONALD W. "Innovation and Academic Research." *Nature,* 316 (August 1, 1985).

BRANDT, RICHARD, et al. "The Future of Silicon Valley." *Business Week,* February 5, 1990.

BUELL, BARBARA. "Philippe Villers, the Social Conscience of Route 128." *Business Week,* February 25, 1985.

*Business Week.* "It's the Morning After for Venture Capitalists." September 24, 1984.

*Business Week.* "New England Highway Upsets Old Way of Life." May 14, 1955.

BRINKMAN, W. F. "A National Engineering and Technology Agency." *Science,* 247 (February 23, 1990).

CAREY, JOHN. "Can U.S. Defense Labs Beat Missiles Into Microchips?" *Business Week,* September 17, 1990.

CAREY, JOHN. "Spurring High Tech on a Shoestring." *Business Week,* August 6, 1990.

CLOGSTON, A. M. "Applied Research: Key to Innovation." *Science,* 235 (January 2, 1987).

DE CASTRO, PAM. "Price Waterhouse Courts High-Tech Entrepreneurs." *Boston Business Journal,* July 7, 1986.

*The Economist.* "The Boom That Went Away." May 5, 1990.

*The Economist.* "How Grey Is My Valley?" March 23, 1991.

ENGSTROM, THERESA. "The Digital Mystique." *Boston Magazine,* August 1981.

FARRELL, KEVIN. "High-Tech Highways." *Venture,* September 1983.

FERGUSON, CHARLES H. "From the People Who Brought You Voodoo Economics." *Harvard Business Review,* May/June 1988.

FIEDLER, TERRY G. "Angels Give Wing to Entrepreneurs." *New England Business,* December 1, 1986.

FOCER, ADA. "The Grand Old Man of Boston Real Estate." *Boston Business,* April/May 1988.

FORD, DAVID, and CHRIS RYAN. "Taking Technology to Market." *Harvard Business Review,* March/April 1981.

*Fortune.* "Business in Isotopes." December 1947.

GLAZER, SARAH. "Businesses Take Root in University Parks." *High Technology,* January 1986.

GOLDMAN, MARSHALL I. "Building a Mecca for High Technology." *Technology Review,* May/June 1984.

HAMILTON, JOAN O'C. "The Big Money Chasing Biotech Brainstorms." *Business Week,* May 16, 1988.

HECTOR, GARY. "A Tough Slog for Venture Capitalists." *Fortune,* June 10, 1985.

HELM, LESLIE. "This Whiz Kid Isn't Such a Whiz at Business." *Business Week,* January 18, 1988.

HOLDEN, CONSTANCE. "Wanted: 675,000 Future Scientists and Engineers." *Science,* 244 (June 30, 1989).

HOUNSHELL, DAVID A., and JOHN KENLY SMITH, JR. "The Nylon Drama." *American Heritage of Invention and Technology,* Fall 1988.

JAHNKE, ART. "Floppy Disks and Floppy Gloves." *Boston Magazine,* July 1985.

JERESKI, LAURA. "Too Much Money, Too Few Deals." *Forbes,* March 7, 1988.

JONES, MEGAN. "Helping States Help Themselves." *Issues in Science and Technology,* Fall 1989.

KEYWORTH, G. A., II. "Federal R&D and Industrial Policy." *Science,* 220 (June 10, 1983).

KIELY, THOMAS, and SUSAN STEINWAY. "The Richest People in Massachusetts." *Boston Magazine,* April 1989.

KNOTT, LEONARD L. "New Blood for New Business." *Canadian Business,* November 1947.

LANDAU, RALPH, and NATHAN ROSENBERG. "America's High-Tech Triumph." *Invention and Technology,* Fall 1990.

LEONE, ROBERT A., and JOHN R. MEYER. "Can the Northeast Rise Again?" *The Wharton Magazine,* 3 [2 (Winter 1979)].

LEVINE, JONATHAN B. "For Venture Capitalists, Too Much of a Good Thing." *Business Week,* June 6, 1988.

LEVINE, JONATHAN B. "Now 'Nontech' Is the Darling of Venture Capitalists." *Business Week,* March 31, 1986.

LEWIS, GEOFFREY C. "Suddenly, Start-ups Are Going Begging." *Business Week,* March 18, 1985.

LINCOLN, FREEMAN. "After the Cabots—Jerry Blakeley." *Fortune,* November 1960.

MAGNUSSON, PAUL. "Promoting High-Definition TV: The Perils for Uncle Sam." *Business Week,* May 29, 1989.

MALECKI, EDWARD J. "Hope or Hyperbole? High Tech and Economic Development." *Technology Review,* October 1987.

MACDONALD, DOUGALD. "Incubator Fever." *New England Business,* September 2, 1985.

MAIN, JEREMY. "Business Goes to College for a Brain Gain." *Fortune,* March 16, 1987.

MARSHALL, ELIOT. "U.S. Technology Strategy Emerges." *Science,* 252 (April 5, 1991).

MASSEY, WALTER E. "Science Education in the United States: What the Scientific Community Can Do." *Science,* 245 (September 1, 1989).

MCLAUGHLIN, MARK. "Defense Spending in Bay State Likely to Be Flat Over Next Five Years." *New England Business,* July 4, 1988.

MCLAUGHLIN, MARK. "Lawrence: From Riots to Redevelopment." *New England Business,* March 3, 1986.

MCLAUGHLIN, MARK. "Out of the Lab, Onto the Market." *New England Business,* July 7, 1986.

MCWILLIAMS, GARY. "The Sons of Apollo May Outshine the Original." *Business Week,* February 18, 1991.

MCWILLIAMS, GARY, et al. "Less Gas for the Bunsen Burners." *Business Week,* May 20, 1991.

MILLER, ROGER, and MARCEL COTE. "Growing the Next Silicon Valley." *Harvard Business Review,* July/August 1985.

NELSON, RICHARD R., and RICHARD N. LANGLOIS. "Industrial Innovation Policy: Lessons from American History." *Science,* 219 (February 18, 1983).

OSBORNE, DAVID. "Refining State Technology Programs." *Issues in Science and Technology,* Summer 1990.

PALCA, JOSEPH. "NSF Centers Rise Above the Storm." *Science,* 251 (January 4, 1991).

PENNAR, KAREN, et al. "The Peace Economy." *Business Week,* December 11, 1989.

PETERS, DONALD H., and E. B. ROBERTS. "Unutilized Ideas in University Laboratories." *Academy of Management Journal,* June 1969.

REICH, ROBERT B. "Who Is Us?" *Harvard Business Review,* January/February 1990.

ROBERTS, EDWARD B. "A Basic Study of Innovators; How to Keep and Capitalize on Their Talents." *Research Management,* XI [4, (1968)].

ROBERTS, EDWARD B., and D. H. PETERS. "Commercial Innovation from University Faculty." *Research Policy,* 10 [2 (April 1981)].

ROBERTS, EDWARD B., and H. A. WAINER. "New Enterprises on Route 128." *Science Journal,* December 1968.

ROBERTS, EDWARD B., and H. A. WAINER. "Some Characteristics of Technical Entrepreneurs." *IEEE Transactions on Engineering Management,* EM-18 [3 (August 1971)].

SALKIND, MICHAEL. "University–Industry Cooperation Is Key to Bolstering Technology Base." *Aviation Week and Space Technology,* December 5, 1988.

SERVOS, JOHN W. "The Industrial Relations of Science: Chemical Engineering at MIT, 1900–1939." *ISIS,* 71 [259 (1980)].

SIMON, JANE. "Route 128: How It Developed, And Why It's Not Likely to Be Duplicated." *New England Business,* July 1, 1985.

SLUTSKER, GARY. "Some Call it Restructuring." *Forbes,* September 16, 1985.

SMITH, KENNETH A. "Industry–University Research Programs." *Physics Today,* February 1984.

Solo, Robert A. "Gearing Military R&D to Economic Growth." *Harvard Business Review,* November/December 1962.

Walker, Francis A. "The Place of the Schools of Technology in American Education." *Educational Review,* October 1891.

Walker, William H. "Chemical Research and Industrial Progress." *Scientific American Supplement,* July 1, 1911.

White, Robert M. "The Crisis in Science Funding: An Interview with Robert M. White." *Technology Review,* May/June 1991.

## MISCELLANEOUS

Appleton, Nathan. "Introduction of the Power Loom, and Origin of Lowell." Lowell, Mass., 1858.

Baty, Gordon. "Initial Financing of the New Research-Based Enterprise in New England." Federal Reserve Bank of Boston, Research Report No. 25, 1964.

Benson, Douglas K. *Sources of Spin-Off Enterprises.* Unpublished S.M. Thesis, MIT Sloan School of Management, 1969.

Birch, David L., Elizabeth A. Martin, and Peter M. Allaman. "Economic Development in Massachusetts in the Early 1970s: An Analysis of Individual Firm Behavior." Joint Center for Urban Studies of MIT and Harvard University, Working Paper No. 6, August 1975.

Bullock, Matthew. "Academic Enterprise, Industrial Innovation, and the Development of High Technology Financing in the United States." London: Brand Brothers and Company, 1983.

Burke, Mary V. "Slow R&D Spending Growth Continues into 1990s." National Science Foundation Report, NSF 90-307, March 9, 1990.

Bush, Vannevar. *Science—The Endless Frontier: A Report to the President on a Program for Postwar Scientific Research.* July 1945. Reprinted May 1980, National Science Foundation.

Carlson, W. Bernard. "Academic Entrepreneurship at MIT: Dugald C. Jackson and the Rise of the Electrical Engineering Department, 1907–1930." ASEE Annual Conference Proceedings, Session 2262, 1984.

COHEN, HOWARD ALLEN. *Spin-Off Organizations: A Study of Enterprises Spun-Off from the M.I.T. Community.* Unpublished S.M. Thesis, MIT Sloan School of Management, 1970.

COMPTON, KARL T. "Relation between Engineering Educators and Industry." Synopsis of address before the General Session of the Society for the Promotion of Engineering Education, June 30, 1937.

DAVIDSON, DEKKERS L. "Massachusetts High Technology Council (A)." Harvard Business School, Case No. 0-383-026, 1982.

DAVIDSON, DEKKERS L. "Massachusetts High Technology Council (B)." Harvard Business School, Case No. 9-383-027, 1982.

DORFMAN, NANCY S. "Massachusetts's High Technology Boom in Perspective: An Investigation of its Dimensions, Causes, and of the Role of New Firms." MIT Center for Policy Alternatives, Report No. CPA 82-2, April 1982.

DUKAKIS, MICHAEL S., and ALDEN S. RAINE. "Creating the Future: Opportunity, Innovation, and Growth in the Massachusetts Economy." State Publication 14906-91-100-7-87-C.R., 1987.

FORSETH, DEAN A. *The Role of Government-Sponsored Research Laboratories in the Generation of New Enterprises—A Comparative Analysis.* Unpublished S.M. Thesis, MIT Sloan School of Management, 1966.

FOSTER, LEROY F. *Sponsored Research at M.I.T.* Cambridge: Division of Sponsored Research, 1984.

GOLDSTEIN, JEROME. *The Spin-Off of New Enterprises from a Large Government-Funded Industrial Laboratory.* Unpublished S.M. Thesis, MIT Sloan School of Management, 1967.

THE GOVERNMENT–UNIVERSITY–INDUSTRY RESEARCH ROUNDTABLE. "Industrial Perspectives on Innovation and Interactions with Universities." Summary of interviews with senior industrial officials, February 1991.

THE GOVERNMENT–UNIVERSITY–INDUSTRY RESEARCH ROUNDTA-

BLE. "Science and Technology in the Academic Enterprise: Status, Trends, and Issues." A discussion paper, October 1989.

HARRINGTON, PAUL E., and ANDREW SUM. "The Economic Recession in Massachusetts: Its Impacts on Labor Markets and Workers Through June 1991." Center for Labor Market Studies, Northeastern University, July 8, 1991.

HARRISON, BENNET. "Rationalization, Restructuring, and Industrial Reorganization in Older Regions: The Economic Transformation of New England Since World War II." Harvard/MIT Joint Center for Urban Studies, Working Paper No. 72, February 1982.

HEKMAN, JOHN S. "New England's High Technology Industry Is Here to Stay." *New England Economic Indicators,* Federal Reserve Bank of Boston, March 1980.

HEKMAN, JOHN S. "The Product Cycle and New England Textiles in the Nineteenth Century." Boston College and Harvard/MIT Joint Center for Urban Studies, May 1978.

HEKMAN, JOHN S. "What Attracts Industry to New England." Federal Reserve Bank of Boston, December 1978.

HOTTEL, HOYT C. "Chemical Engineering at MIT in the '20s." Speech delivered at the MIT Centennial of Chemical Engineering, October 7, 1988.

JOINT BOARD FOR THE METROPOLITAN MASTER HIGHWAY PLAN. *Master Highway Plan,* 1948.

LAMPE, DAVID R. "MIT's Role in the Development of the Boston Area," in *Die Zukunft der Metropolen: Paris, London, New York, Berlin,* Vol. 1. Technical University of Berlin, August 1984.

LAMPE, DAVID R., and JAMES M. UTTERBACK. "The Tradition of University–Industry Relations at MIT." Paper presented at the International Conference on University–Industry Innovation, Milano, Italy, October 1983.

LEDERMAN, LEON M. "Science: The End of the Frontier?" A Report to the Board of Directors of the American Association for the Advancement of Science, January 1991.

MACDONALD, ALLAN, and EDWARD KAZNOCHA. "Defense Spending

and Massachusetts Employment, 1972–1980." Commonwealth of Massachusetts Division of Employment Security, October 1982.

MASSACHUSETTS INSTITUTE OF TECHNOLOGY. Unpublished study that accompanied the release of MIT commemorative publication, "Event 128: A Salute to Founders." April 1989.

MOORE, CRAIG L., and STEVEN M. ROSENTHAL. *Massachusetts Reconsidered: An Economic Anatomy of the Commonwealth.* School of Business Administration, University of Massachusetts at Amherst, 1978.

NATIONAL SCIENCE BOARD. "University–Industry Relationships: Myths, Realities, and Potentials." 14th Annual Report of the National Science Board, National Science Foundation, Washington, D.C., 1982.

NEW ENGLAND BOARD OF HIGHER EDUCATION. "Biomedical Research and Technology: A Prognosis for International Economic Leadership." Commission on Academic Medical Centers and the Economy of New England, June 1988.

NEW ENGLAND BOARD OF HIGHER EDUCATION. "A Threat to Excellence." Preliminary Report of the Commission on Higher Education and the Economy of New England, March 1982.

NEW ENGLAND COUNCIL, INC. "Education: Cornerstone of New England's Prosperity." Proceedings of the 61st Annual Conference, November 20, 1986.

OLSEN, KENNETH H. "Digital Equipment Corporation: The First Twenty-five Years." Address to The Newcomen Society in North America, Boston, September 21, 1982.

PORTER, MICHAEL E. "The Competitive Advantage of Massachusetts." Independent Report. 1991.

RUBENSTEIN, ALBERT H. "Problems of Financing and Managing New Research-Based Enterprises in New England." Federal Reserve Bank of Boston, Research Report No. 3, 1958.

SAHLMAN, WILLIAM A., and HOWARD H. STEVENSON. "Capital Market Myopia." Harvard Business School Note, 1987.

SINCLAIR, DONALD B. "The General Radio Company, 1915–1965." Address to The Newcomen Society in North America, Boston, May 13, 1965.

SIRBU, MARVIN A., et al. "The Formation of a Technology Oriented Complex: Lessons from North American and European Experience." MIT Center for Policy Alternatives, Report No. CPA 76-8, December 1976.

STAROBIN, PAUL. "Business Lobbying in the Bay State: The Massachusetts High Technology Council." John F. Kennedy School of Government, Harvard University, Case No. C14-83-500, 1983.

STAROBIN, PAUL. "Business Lobbying in the Bay State: The Massachusetts High Technology Council: Sequel." John F. Kennedy School of Government, Harvard University, Case No. C14-83-500S, 1983.

TAYLOR, CHRISTOPHER L. *New Enterprises Descended from a Technically Based Company.* Unpublished S.M. Thesis, MIT Sloan School of Management, 1981.

TEPLITZ, PAUL V. *Spin-Off Enterprises from a Large Government-Sponsored Laboratory.* Unpublished S.M. Thesis, MIT Sloan School of Management, 1965.

THIESSEN, ARTHUR E. *A History of the General Radio Company.* West Concord, Massachusetts, 1965.

WAINER, HERBERT A. *The Spin-Off of Technology from Government-Sponsored Research Laboratories: Lincoln Laboratory.* Unpublished S.M. Thesis, MIT Sloan School of Management, 1965.

ZIEGLER, CHARLES ALBERT. *Looking Glass Houses: A Study of the Process of Fissioning in an Innovative, Science-Based Firm.* Unpublished Ph.D. Dissertation, Brandeis University, 1982.

## NEWSPAPER ARTICLES

ADAMS, JANE MEREDITH. "In Trouble, Ovation Looks for a Buyer." *The Boston Globe,* October 16, 1984.

ANDERSON, PETER. "Can't Live With It, Can't Live Without It: Along the Not-So-Lonesome Highway Called 128." *The Boston Globe Magazine,* January 12, 1986.

ASHBROOK, TOM. "Biogen $57.5M Stock Offering a Sellout." *The Boston Globe,* March 23, 1983.

BEAM, ALEX. "Zorched Out: A Computer Hacker's Tale." *The Boston Globe,* February 7, 1988.

BECKER, RENE. "Mr. Wizard: The Inventive World of Raymond Kurzweil." *The Boston Globe Magazine,* June 5, 1988.

BLUESTONE, BARRY, and BENNETT HARRISON. "A Low-Wage Explosion: The Grim Truth about the Job 'Miracle.' " *The New York Times,* February 1, 1987.

*The Boston Globe.* "Osborne Computer Files Chap. 11." September 15, 1983.

BULKELEY, WILLIAM M. "Computer Engineers Memorialized in Book Seek New Challenges." *The Wall Street Journal,* September 20, 1985.

BULKELEY, WILLIAM M. "Encore's Fisher Gets Another Chance to Make It Big." *The Wall Street Journal,* April 11, 1989.

BUTTERFIELD, BRUCE. "The Massachusetts Miracle Is Over." *The Boston Globe,* February 11, 1989.

BUTTERFIELD, FOX. "High-Tech Plugging into New England." *The New York Times,* July 8, 1984.

CURWOOD, STEVE. "Biotech Bellyache." *The Boston Globe,* August 23, 1983.

DIMANCESCU, DAN. "Centers of Excellence: Reality or Illusion?" *The Boston Globe,* November 19, 1985.

EDELMAN, LAWRENCE. "Bull HN to Lay Off 1,000 in Bay State." *The Boston Globe,* November 9, 1990.

EDELMAN, LAWRENCE. "Poduska Named CEO of Stardent." *The Boston Globe,* October 31, 1989.

EDELMAN, LAWRENCE. "Route 128 Awakening to a Brave 'New-Tech' World." *The Boston Globe,* June 30, 1991.

ELLIOTT, J. RICHARD, JR. "Something Ventured." *Barron's,* March 6, 1967.

FOX, WENDY. "A New Tune from Lotus." *The Boston Globe,* July 1, 1984.

FRENCH, DESIREE. "Data General Head Criticizes Dukakis on State Economy." *The Boston Globe,* January 18, 1984.

FULMAN, DIANE. "Economic Adversity Sows Seeds of Future Business Opportunity." *The Boston Globe,* March 20, 1990.

GARBODEN, CLIFF. "Primer: Stanley Steamer." *The Boston Globe Magazine,* June 25, 1989.

GEISER, KENNETH, and BENNETT HARRISON. "The High-Tech Industry Comes Down to Earth." *The Boston Globe,* June 23, 1985.

GOROV, LYNDA. "Automatix Co-Founder Quits." *The Boston Globe,* March 20, 1986.

GOSSELIN, PETER. "Dollars for Science." *The Boston Globe,* September 10, 1989.

GRAHAM, RENEE. "Route 128 Choking on Prosperity." *The Boston Globe,* June 15, 1988.

GUPTA, UDAYAN. "How an Ivory Tower Turns Research into Start-Ups." *The Wall Street Journal,* September 19, 1989.

KAPLAN, FRED. "Massachusetts Organizations Get $700M in 'Star Wars' Money." *The Boston Globe,* October 28, 1986.

KENNEY, CHARLES. "Boston Future." *The Boston Globe,* June 19, 1988.

KLEINFIELD, N. R. "A Few Clouds Over Route 128." *The New York Times,* July 18, 1984.

LENZNER, ROBERT. "Computer Stocks—Big and Small—Taking a Beating." *The Boston Globe,* June 19, 1984.

MOHL, BRUCE. "Dukakis Concedes 'Miracle' Is Gone." *The Boston Globe,* November 15, 1989.

MOHL, BRUCE. "State Hits Bottom in Bond Rating." *The Boston Globe,* November 16, 1989.

PATTERSON, GREGORY A. "Defense Industry Sees Hard Times Ahead." *The Boston Globe,* November 22, 1987.

PRESS, FRANK. "Science Shouldn't Get Low Priority, Even in This Era of Deficits." *The Boston Globe,* May 5, 1991.

ROSENBERG, RONALD. "AI Alley's Longest Winter." *The Boston Globe,* December 18, 1988.

ROSENBERG, RONALD. "The Boom Is on Along Route 128." *The Boston Globe,* February 7, 1984.

ROSENBERG, RONALD. "Encore Computer Unveils Plans." *The Boston Globe,* July 22, 1983.

ROSENBERG, RONALD. "A Firm Goes Public, a Millionaire Is Made." *The Boston Globe,* October 7, 1983.

ROSENBERG, RONALD. "The High Rollers of Venture Capital Hit the Wall." *The Boston Globe,* November 17, 1985.

ROSENBERG, RONALD. "Intelligence Alley." *The Boston Globe,* August 6, 1985.

ROSENBERG, RONALD. "Knoware, Educational Software Firm, Is Broke." *The Boston Globe,* October 17, 1984.

ROSENBERG, RONALD. "Playing It Safe." *The Boston Globe,* September 4, 1988.

ROSENBERG, RONALD. "Six Defect to Join Ex-Prime Boss." *The Boston Globe,* July 16, 1983.

ROSENBERG, RONALD. "Slowdown Snags Software." *The Boston Globe,* May 12, 1985.

ROSENBERG, RONALD. "Tough Times Ahead for Venture Capitalists." *The Boston Globe,* December 9, 1984.

ROSENBERG, RONALD. "Venture with Care." *The Boston Globe,* October 28, 1986.

ROSENBERG, RONALD. "William Poduska's Hat Trick." *The Boston Globe,* April 14, 1987.

SCHRAGE, MICHAEL. "Venture Capital's New Look." *The Wall Street Journal,* May 19, 1988.

SEGE, IRENE. "All Signs Point to Mass Exodus." *The Boston Globe,* July 23, 1991.

SIMON, JANE FITZ. "Apollo: From Star to Also-Ran." *The Boston Globe,* April 13, 1989.

SIMON, JANE FITZ. "Apple Plans Laboratory in Cambridge." *The Boston Globe,* January 5, 1989.

SIMON, JANE FITZ. "Cognition Inc., Citing Lack of Capital, Seeks Buyer." *The Boston Globe,* January 4, 1989.

SIMON, JANE FITZ. "Is Biotech the Savior?" *The Boston Globe,* December 12, 1989.

SIMON, JANE FITZ. "Philippe Villers Quits as Cognition President." *The Boston Globe,* December 9, 1988.

SIMON, JANE FITZ. "Software, Made Simple: A Fresh Quest for Kapor." *The Boston Globe,* November 17, 1987.

SIMON, JANE FITZ. "Stellar Is Set to Announce Merger with Rival Ardent." *The Boston Globe,* August 30, 1989.

SIMON, JANE FITZ. "Three Buck the Tide." *The Boston Globe,* September 19, 1989.

SIMON, JANE FITZ. "Top Officers Step Down at Symbolics." *The Boston Globe,* February 5, 1988.

STEIN, CHARLES. "Defense Boom to Continue Through 1987." *The Boston Globe,* January 13, 1987.

STEIN, CHARLES. "High Technology's Promising New Wave." *The Boston Globe,* June 11, 1991.

STEIN, CHARLES. "Living in the Shadow of the '80s." *The Boston Globe,* December 30, 1990.

STEIN, CHARLES. "The Rise and the Fall." *The Boston Globe,* January 30, 1990.

STEIN, CHARLES. "Still Cranky after 10 Years." *The Boston Globe,* September 13, 1987.

STEIN, CHARLES. "Villers to Develop a Third Company." *The Boston Globe,* January 16, 1985.

STEIN, CHARLES. "When Bad Things Happen to Good Governors." *The Boston Globe,* April 1, 1990.

TOLCHIN, MARTIN. "Crucial Technologies: 22 Make the U.S. List." *The New York Times,* March 17, 1989.

WARSH, DAVID. "Bouncing Back." *The Boston Globe,* September 3, 1991.

WARSH, DAVID. "Some Miracles Never Cease." *The Boston Globe Magazine,* May 7, 1989.

WESSEL, DAVID. "Computer Devices Petitions for Protection." *The Boston Globe,* November 1, 1983.

WESSEL, DAVID. "Data General's Progeny." *The Boston Globe,* September 6, 1983.

WESSEL, DAVID. "Health-Care Dropout Cashes in for $70M." *The Boston Globe,* October 11, 1983.

WESSEL, DAVID. "The Making of Knoware." *The Boston Globe,* September 23, 1983.

WILKE, JOHN. "Apollo Agrees to Be Bought for $476M." *The Boston Globe,* April 13, 1989.

WILKE, JOHN. "Biotech Comes into Its Own." *The Boston Globe,* October 21, 1986.

WILKE, JOHN. "Data General's Fight for Life." *The Boston Globe,* April 25, 1989.

WILKE, JOHN. "Prices, Politics, Climate Cited in Loss of Sematech." *The Boston Globe,* January 7, 1988.

# Index

# INDEX